Springer
Proceedings in Physics 80

Springer
Berlin
Heidelberg
New York
Barcelona
Budapest
Hong Kong
London
Milan
Paris
Santa Clara
Singapore
Tokyo

Springer Proceedings in Physics

Managing Editor: H. K. V. Lotsch

46 *Cellular Automata and Modeling of Complex Physical Systems*
Editors: P. Manneville, N. Boccara, G. Y. Vichniac, and R. Bidaux

47 *Number Theory and Physics*
Editors: J.-M. Luck, P. Moussa, and M. Waldschmidt

48 *Many-Atom Interactions in Solids*
Editors: R. M. Nieminen, M. J. Puska, and M. J. Manninen

49 *Ultrafast Phenomena in Spectroscopy*
Editors: E. Klose and B. Wilhelmi

50 *Magnetic Properties of Low-Dimensional Systems II: New Developments*
Editors: L. M. Falicov, F. Mejía-Lira, and J. L. Morán-López

51 *The Physics and Chemistry of Organic Superconductors*
Editors: G. Saito and S. Kagoshima

52 *Dynamics and Patterns in Complex Fluids: New Aspects of the Physics–Chemistry Interface*
Editors: A. Onuki and K. Kawasaki

53 *Computer Simulation Studies in Condensed-Matter Physics III*
Editors: D. P. Landau, K. K. Mon, and H.-B. Schüttler

54 *Polycrystalline Semiconductors II*
Editors: J. H. Werner and H. P. Strunk

55 *Nonlinear Dynamics and Quantum Phenomena in Optical Systems*
Editors: R. Vilaseca and R. Corbalán

56 *Amorphous and Crystalline Silicon Carbide III, and Other Group IV-IV Materials*
Editors: G. L. Harris, M. G. Spencer, and C. Y. Yang

57 *Evolutionary Trends in the Physical Sciences*
Editors: M. Suzuki and R. Kubo

58 *New Trends in Nuclear Collective Dynamics*
Editors: Y. Abe, H. Horiuchi, and K. Matsuyanagi

59 *Exotic Atoms in Condensed Matter*
Editors: G. Benedek and H. Schneuwly

60 *The Physics and Chemistry of Oxide Superconductors*
Editors: Y. Iye and H. Yasuoka

61 *Surface X-Ray and Neutron Scattering*
Editors: H. Zabel and I. K. Robinson

62 *Surface Science: Lectures on Basic Concepts and Applications*
Editors: F. A. Ponce and M. Cardona

63 *Coherent Raman Spectroscopy: Recent Advances* Editors: G. Marowsky and V. V. Smirnov

64 *Superconducting Devices and Their Applications*
Editors: H. Koch and H. Lübbig

65 *Present and Future of High-Energy Physics*
Editors. K.-I. Aoki and M. Kobayashi

66 *The Structure and Conformation of Amphiphilic Membranes*
Editors: R. Lipowsky, D. Richter, and K. Kremer

67 *Nonlinearity with Disorder*
Editors: F. Abdullaev, A. R. Bishop, and S. Pnevmatikos

68 *Time-Resolved Vibrational Spectroscopy V*
Editor: H. Takahashi

69 *Evolution of Dynamical Structures in Complex Systems*
Editors: R. Friedrich and A. Wunderlin

70 *Computational Approaches in Condensed-Matter Physics*
Editors: S. Miyashita, M. Imada, and H. Takayama

71 *Amorphous and Crystalline Silicon Carbide IV*
Editors: C. Y. Yang, M. M. Rahman, and G. L. Harris

72 *Computer Simulation Studies in Condensed-Matter Physics IV*
Editors: D. P. Landau, K. K. Mon, and H.-B. Schüttler

73 *Surface Science: Principles and Applications*
Editors: R. F. Howe, R. N. Lamb, and K. Wandelt

74 *Time-Resolved Vibrational Spectroscopy VI*
Editors: A. Lau, F. Siebert, and W. Werncke

75 *Computer Simulation Studies in Condensed-Matter Physics V*
Editors: D. P. Landau, K. K. Mon, and H.-B. Schüttler

76 *Computer Simulation Studies in Condensed-Matter Physics VI*
Editors: D. P. Landau, K. K. Mon, and H.-B. Schüttler

77 *Quantum Optics VI*
Editors: D. F. Walls and J. D. Harvey

78 *Computer Simulation Studies in Condensed-Matter Physics VII*
Editors: D. P. Landau, K. K. Mon, and H.-B. Schüttler

79 *Nonlinear Dynamics and Pattern Formation in Semiconductors and Devices*
Editor: F.-J. Niedernostheide

80 *Computer Simulation Studies in Condensed-Matter Physics VIII*
Editors: D. P. Landau, K. K. Mon, and H.-B. Schüttler

Volumes 1–45 are listed at the end of the book

D. P. Landau K. K. Mon H.-B. Schüttler (Eds.)

Computer Simulation Studies in Condensed-Matter Physics VIII

Recent Developments

Proceedings of the Eighth Workshop
Athens, GA, USA, February 20–24, 1995

With 88 Figures

Springer

Professor David P. Landau, Ph. D.
Professor K. K. Mon, Ph. D.
Professor Heinz-Bernd Schüttler, Ph. D.

Center for Simulational Physics
The University of Georgia
Athens, GA 30602, USA

Cataloging-in-Publication Data applied for

Die Deutsche Bibliothek - CIP-Einheitsaufnahme

Computer simulation studies in condensed matter physics : proceedings of the ... workshop ... - Berlin ; Heidelberg ; New York ; London ; Paris ; Tokyo ; Hong Kong ; Barcelona ; Budapest : Springer.

8. Athens, GA, USA, February 20 - 24, 1995. - 1995
(Springer proceedings in physics ; 80)
ISBN-13:978-3-642-79993-8 e-ISBN-13:978-3-642-79991-4
DOI: 10.1007/978-3-642-79991-4
NE: GT

ISBN-13:978-3-642-79993-8

Softcover reprint of the hardcover 1st edition 1995

Typesetting: Camera ready copy from the authors/editors
SPIN: 10502525 54/3144 - 5 4 3 2 1 0 - Printed on acid-free paper

Preface

The role computer simulation is playing in both research and instruction in many areas of physics has been growing quite rapidly. As the simulations community began to expand, we sensed a need for a meeting place for experienced simulators and neophytes to discuss new techniques and recent results in an environment which promoted extended discussion. As a consequence, the Center for Simulational Physics established an annual workshop on *Recent Developments in Computer Simulation Studies in Condensed-Matter Physics*. This year's workshop was the eighth in this series, and the interest shown by the scientific community demonstrates quite clearly the useful purpose the series has served. The latest workshop was held at The University of Georgia, February 20–24, 1995, and these proceedings form a record that is published with the goal of timely dissemination of the material to a wider audience.

This volume is composed of three parts. The first part contains invited papers that deal with simulational studies of classical systems. The second part is devoted to invited papers on quantum systems, including new results for strongly correlated electron and quantum spin models. The final part comprises contributed presentations.

We hope that each reader will benefit from specialized results as well as profit from exposure to new algorithms, methods of analysis, and conceptual developments. We have already learned that fruitful collaborations and new research projects have resulted from these gatherings at The University of Georgia, and we hope that these proceedings will further expand this tendency.

Athens, GA, USA
June 1995

D.P. Landau
K.K. Mon
H.-B. Schüttler

Contents

Part III Contributed Papers

An Introduction

D.P. Landau, K.K. Mon and H.-B. Schüttler

Center for Simulational Physics, The University of Georgia, Athens, GA 30602-2451, USA

The 8th Annual Workshop featured presentations on a broad range of topics. These proceedings are divided into three parts. The first two parts contain papers pertaining to invited presentations and contain pedagogical material. The last section contains shorter, contributed papers.

To open Part I, J. Adler presents an overview of modern methods for generation and interactive analysis of series expansions for the study of critical behavior. For comparison, results for several different models for which Monte Carlo results also exist are presented. The relative advantages of and resources required for each method are presented. N. B. Wilding then considers application of finite-size-scaling techniques to the simulation of critical fluids. The author presents a theory which can account for the broken symmetry between the coexisting phases of fluids and the presence of field mixing. The theoretical predictions are tested with extensive Monte Carlo study of the critical density and energy fluctuations of the three-dimensional Lennard-Jones fluid. The results demonstrate that highly accurate determinations of fluid critical point parameters are possible. In the next paper, P. Smilauer and D. D. Vvedensky describe the use of kinetic Monte Carlo simulations to study epitaxial growth and chemical beam etching of GaAs(001). Direct comparison between measured RHEED intensity and the simulated surface step density on a GaAs(001) vicinal surface shows excellent agreement. Growth and etching of singular surfaces are also modeled. M. L. Plumer and A. Mailhot review their recent finite-size-scaling results of Monte Carlo data for the stacked triangular antiferromagnetic Ising, XY and Heisenberg models. The authors conclude that the Ising case is in the XY universality class. The stacked XY model data supports a tricritical behavior. A new n=3 chiral universality class is consistent with the results for the stacked Heisenberg model. To complete Part I, D. Wolf presents a simple absolutely convergent solution of the Madelung problem. The approach is very interesting and is based on summing over neutral shells of the Bravais lattice. This produced an effective Coulomb potential in ionic crystals and liquids which decreases as $1/r^5$. The insights from this technique lead to understanding of ionic-crystal surface reconstruction and polar surfaces.

In the first paper in the section on quantum systems, A. Omeltchenko et al. report on large-scale molecular dynamics simulations of amorphous materials, using multi-resolution algorithms which have been implemented on massively parallel machines. Detailed results for the thermal transport properties of amorphous silicon nitride are presented. S. Miyazawa and S. Homma report on a Monte Carlo study of the frustrated spin-1/2 Heisenberg model with antiferromagnetic 1st and 2nd neighbor couplings, J_1 and J_2, on two-dimensional square lattice, using a modified version of the decoupled cell method. The focus of that paper is the competition between the Neel and collinear ordered states in that model. Lastly, J. J. Deisz et al. report on a new approach for obtaining dynamical spectral functions in the real-frequency domain at zero temperature by combining projection Monte Carlo with maximum entropy techniques.

Computer Simulation Studies in Condensed-Matter Physics VIII
Editors: D.P. Landau · K.K. Mon · H.-B. Schüttler

In the first contributed paper, D. Ledue et al. study an Ising ferromagnet on the quasiperiodic octagonal tiling via Monte Carlo simulations and determine the transition temperature and critical exponents. G. A. Baker, Jr. and N. Kawashima report new Monte Carlo results of the renormalized coupling constant in the three-dimensional Ising model, employing improved estimators. Their results strongly support the hyperscaling hypothesis. K. Vollmayr et al. present molecular dynamics results on cooling rate effects in a Lennard-Jones glass. This work sheds new light on the dynamics and the global and local properties of meta-stable free energy minima near the glass transition. The paper by J. Zhang et al. presents lattice gas model results for the coadsorption of copper and sulfate ions on the (111) surface of gold electrodes in sulfate-containing electrolytes. Using Monte Carlo simulation techniques, they have studied the formation of an ordered $\sqrt{3} \times \sqrt{3}$ mixed surface layer phase as a function of the electrode potential. J. L. Lebowitz et al. use Monte Carlo techniques to study the multicomponent Widom-Rowlinson models (M species and nearest neighbor hard core exclusion between unlike particles) on a square lattice. For M between two and six, a direct transition from the gas phase to a demixed phase is found. For M larger than or equal to seven, the model has an intermediate ordered phase in which sublattice is occupied preferentially. I. Chang et al. report a study of the localization transition in quantum percolation and conclude that the transition in this model belongs to a new universality class. B. V. Costa et al. use spin dynamics simulations to study vortex motion in a 2-dim classical, easy plane Heisenberg model. Above the transition they find that vortex-anti-vortex creation-annihilation dominates. D. C. Rapaport provides a perspective on the use of interactive computer graphics as an aid in research. Design considerations and examples are given K. Eichhorn and K. Binder study the phase transitions of the three-dimensional three-state Potts model in random fields. The results indicate a first-order transition for strong random fields and a second-order transition for intermediate field strength. P. J. Fay et al. present a method of determining the chemical potential in Monte Carlo or molecular dynamics simulations by determining the condition for detailed balance between fictitious creation and destruction attempts. Results for Lennard-Jones model and Palladium model are given. M. Krech and D. P. Landau report Monte Carlo results for the Casimir effect in critical stripes. Using a novel type of hybrid algorithm, they have studied the Casimir effect in the context of a critical q-state Potts model, confined to rectangular geometries. The paper by M. Suzuki introduces a new scheme of exponential product formulae with relevance for quantum Monte Carlo simulations. A general scheme for constructing hybrid product formulae is given and several explicit examples are included. H. G. Evertz and D. P. Landau report results from a spin dynamics study of the classical d=2 xy model. The dynamic structure factor is described and an estimate is given for the dynamic critical exponent.

Part I

Classical Systems

Modern Series Analysis Techniques and the Relation to Monte-Carlo Results on Similar Models

J. Adler

Department of Physics, Technion-IIT, 32000, Haifa, Israel

Abstract. In this paper an introduction to modern techniques of series analysis and a comparison of results with those of Monte Carlo simulations is given. Discussions of the relative advantages of and resources required for both methods and the relative precision of the results are presented.

1. Introduction

This article complements two recent reviews by the author: an introduction [1] to both generation and analysis of series expansions in a percolation context for a student audience, and a comparison [2] of results from the analysis of series with simulations for two exactly solved models (Ising and Baxter-Wu). Both are aimed at a simulation audience and the former includes sample programs. General introductory references to different aspects of series expansion methods can be found in [3-12].

In the present article a comparison of results from series, simulations and some other techniques for several systems will be given. A comparative discussion of the two methods in Section 2, will include a comparison of the relative resources needed, advantages and disadvantages of both and some historical developments. The clear division between series generation and analysis will be described in Section 3, together with an introduction to some methods of analysis that allow for the effects of non-analytic corrections-to-scaling. In Section 4, results for several systems from both simulations and series will be presented. Conclusions pertaining to the relative merits of series and simulations will be given in Section 5.

2. A General Comparison of Series and Simulations

Since the following comparison is written from a series-expansion point-of-view some simulation proponents amongst the readers may find it a little biased.

Springer Proceedings in Physics, Vol. 80
Computer Simulation Studies in Condensed-Matter Physics VIII
Editors: D.P. Landau · K.K. Mon · H.-B. Schüttler

However, even if you do think there is bias in favour of series please bear with in order to appreciate the reasons for the resurgence of interest in series expansions that has begun in few years. Since the following is also written from the viewpoint of Computational Physics (not all supporters of series methods really like to see them categorised as Computational Physics) others may find other biases.

Simulation methods began slowly in the fifties and sixties and have continued to gain momentum ever since as computer technology developed. Simulations are based on studying a selection of realizations of finite systems, and would give exact results if the system were infinite and if a sufficient number of independent configurations could be generated. Overall, the total collection of results from computer simulations is impressive. While there is never enough computer time for any quality simulation, (just ask the committees dividing computer time) any time available can be always be utilised to obtain a few more samples or a better check on equilibration. Simulations are in fact extremely efficient in terms of their utilization of computer resources. However there are some problems with simulations. For example for some systems of interest, such as spin glasses, it can be difficult to determine that the simulated system has reached equilibrium and to extrapolate the results to the thermodynamic limit. Also for problems of current concern to the general condensed matter community, such as groundstates of quantum systems, simulations seem to have problems of principle as to how to simulate to obtain answers to some questions of interest. In addition to these examples of specific problems there is the general fact that one must admit that a simulation is always restricted to a particular model in a given spatial dimension at one time. Thus in order to study the effect of changing temperature more than a little or changing disorder even a little many different calculations need to be made. Simulations tend to be difficult to do in high dimensions and at very low temperatures where there can be long relaxation times. Therefore there is certainly room in modern Computational Physics for alternative numerical techniques. The most developed alternative is the series expansion method, which is especially good for low-temperatures, and high dimensions and has no involvement (and therefore no problems) with random number generators.

In contrast to the smooth development of simulations and the consistent success of simulations in achieving efficient use of avaliable computer resources, the development of series expansions has been more erratic, and incompatible at times with state-of-the art hardware. The series method flowered from 1960 on and provided important input for the ideas of scaling and universality that underlie the renormalization group. However certain problems, that will be discussed in detail in the next paragraph, led to the situation where even the biggest fans of the series approach were forced to grudgingley admit that simulations really were the dominant numerical technique for critical phenomena in the eighties. After presenting the solutions to these problems with the series methods I shall argue that the relative advantage of simulations to series is diminishing as simulations become more complex and series generation and analysis easier.

Series methods have traditionally used less computer resources than simulations. The ideal series machine of the seventies was the mainframe, situated locally with access to a tape drive to read in graph lists and read out components of the series. Preferably, this machine belonged to a small research group. As long as such machines were all one had, and series could use these old mainframes exactly like simulations, the intrinsic efficiency of the series methods led to superior results. However with the move to remote-site vector supercomputers in the eighties insurmountable problems with both vectorization and data transmission of graph lists led to effective limits on resources avaliable for series, and as everyone was jumping on the vectorization bandwagon, series were left behind. Coinciding with this resource question some justified unpopularity for series resulted from apparent violations of hyperscaling, that were caused by inappropriate extrapolation techniques that neglected corrections-to-scaling. These two effects together led to a definite downturn in activity relative to the growth of simulation methods at this time, although some groups remained active in series expansions and laid the foundation for the current revivial. The analysis question is now well under control and while series methods have not been adapted for vector computers in the meantime, other recent developments in computing have been so favourable to series expansions calculations that the matter of vectorisability is largely irrelevant as we shall now show.

Despite my biased(?) view that simulation proponents reponded very well to changes in computer hardware, many simulation experts think that they have always experienced serious problems of insufficient time and memory. As a result of this there has been an incredible advance in development of algorithms for simulations in the last fifteen years. This push began with development of algorithms suited to special purpose and vector computers of the last decade and has continued on the parallel machines more recently. This increased sophistication has a downside because prior to the development of analysis routines of the histogram weighting type, simulations tended to demand very little disk storage or communication overhead and so were ideally suited to the remote site vector supercomputers of the eighties. Now they would experience some of the same limitations in data transmission and need for disk storage as do series methods, if these needs were not now being met by recent advances in computers and communications.

Four recent computing advances that have been most helpful to series expansion studies are:

1. fast communications and networking,
2. widespread use of workstations with large disks and memories,
3. parallel machines, and
4. MATHEMATICA and MAPLE computer algebra packages.

The first three of these have also helped simulations, but relatively speaking improved data transmissions have meant that for series remote-site work has moved from almost impossible unless tapes were sent, to commonplace whereas for simulations it has been a gradual improvement. Computer algebra on the other hand has not had much impact on simulations but has played an enormous rôle in series development for both analysis and generation. At least until recently the time overhead on starting a new series project was far greater than for a simulation, since series algorithms were relatively more complex. This is no longer the case as, at least for startup, computer algebra can simplify the complexity (a few terms can be easily generated with computer algebra, even if it is too slow for an entire project), and simulation algorithms have become more complex.

3. Generation and Analysis of Series

Series expansions are one member of the general family of perturbation expansions around some limit of the model for which an exact solution can be found. In all of these extrapolation must be made to the region of the model that is of interest, and this is not always reliable. For example, in some field-theoretic methods one expands around the upper critical dimension. This can be as high as six or even eight for some interesting problems and hence extrapolation to physical dimensions may be questionable. One perturbation method that has proved extremely successful in general dimension is the method of exact series expansions. Here the exact expression for the order parameter, susceptibility or other desired quantity is obtained as a series in increasing powers of some variable such as inverse temperature, T^{-1}, for magnets, or concentration, p, of conducting particles in a mixture of insulating and conducting particles (the percolation problem). This expansion is calculated on a term-by-term basis, and usually uses graph enumerations. The expansion would give us the exact solution if it could be carried to infinite order, but near the limit of low (high) concentration an expansion in p ($q = 1 - p$) to a finite order will be quite reliable. Extrapolation is made towards the region where the expansion variable reaches a critical value and the quantity being expanded exhibits a singularity. Expansions from the disordered phase are especially well behaved; for magnetic systems this is the high temperature, and for percolation the low-concentration limit. The best results are obtained when extrapolations can be made from both extremes and the results shown to be consistent.

We note that series expansion data are extremely compact for storage, and hence many analyses can easily be made on the same series data. Secondly, the series contain information on all possible thresholds, unlike simulations which, in many cases, must be carried out separately for each trial temperature or threshold value. Thus, if there appears a reason to investigate the behaviour in another location at a later date, it is possible to obtain a new prediction from the series by recalling a very few coefficients rather than repeating the entire calculation.

Series expansions studies are always divided into two parts; generation and analysis. See [1-2] for a recent discussion of this question. The reliability of series generation has an impressive record. While series can give different critical parameters based on the analysis method there is only one correct series for each quantity for each model. Generation with different algorithms has consistently led to identical terms for a wide variety of problems. This is in contrast to the simulation world, where if nothing else random number generation seems to be an endless source of worry. A summary of the current state of algorithms for the generation of series is given in [2] together with a comphrensive list of references. Several of the algorithms (especially those for high temperature/low density series) enable simultaneous generation of series for general dimension and disorder or field parameters in a single calculation.

To analyse a series (or a simulation) one must have some understanding of the critical behaviour and make suitable hypotheses about its form. Analysis of series does seem to be more complex than analysis of simulations (at least prior to the advent of the histogram methods.) For series not at critical dimensions where logarithms complicate matters, critical behaviour is usally of the form (on the low temperature side for the magnetization, M)

$$M \sim (T_c - T)^{\beta}(1 + A(T_c - T)^{\Delta_1} + ...) \tag{4}$$

Here T_c is the critical temperature, β is a dominant critical exponent and Δ_1 a "correction-to-scaling" exponent. This is also the exact form for the Baxter-Wu model. In the two-dimensional Ising case Δ_1 is simply analytic (this is related to non-linear scaling fields), and so

$$M \sim (T_c - T)^{\beta}(1 + A(T_c - T) + ...), \tag{3}$$

The two-dimensional Ising is the exception and a non-analytic "correction-to-scaling" such as observed in the Baxter-Wu model is typical of most models of physical interest (including the three-dimensional Ising). Its origin is usually additional operators that are irrelevant in the renormalization group sense. These "corrections-to-scaling" are of paramount importance to series analysis. Failure to allow for their effect causes exponent estimates that violate thermodynamic relations known as hyperscaling relations. Many of the results presented in the next section are based on two methods developed by the author in collaboration with M. Moshe and V. Privman [7,8] in the early eighties that explicitly allow for corrections-to scaling. Other good methods are discussed in [5]. Our methods are based on transforming the series to minimize the interference of the correction terms prior to calculating the approximants. They are variants of the threshold-biased dlog Padé, and one has been based on a generalization of the Roskies transform. Here a series in the variable T (which could stand for p too) is transformed to one in the variable $y = 1-(1-T/T_c)^{\Delta_1}$. We obtain Padé approximants to the series $G(y) = \Delta_1(y-1)\mathrm{d}(\ln M(T))/\mathrm{d}T$. At the correct value of Δ_1, a set of different high and near-diagonal threshold-biased approximants of $G(y)$ all should give the correct dominant exponent, for example β if we study magnetization. It can be shown [8] that deviations from the correct value will be seen as changes in the slope of the dominat estimates plotted as a function of the input value of Δ_1. Such a plot will have an intersection region near the correct (Δ_1, β) point. A related method involves calculating the logarithmic derivative of $B(T) = \gamma M(T) - (T_c - T)\, d(M(T))/dT$ which has a pole at T_c with residue $\beta - \Delta_1$. Again threshold-biased approximants are calculated and graphed and an intersection region is found near the correct (Δ_1, β) point. In many papers these methods are known as M2 and M1, respectively. Interactive graphics subroutines are very useful for these methods and were described in Ref. [7]. If the value of T_c is not known, different trial temperature values are tested and convergence in the three-dimensional T_c, β, Δ_1 space is sought. If we attempt to draw graphs at trial T_c values that differ from the exact ones then we can clearly illustrate the optimal convergence in the correct T_c plane. An example of such a graph is shown for the susceptibilty of five-dimensional bond percolation for the M2 analysis in Section 4.4. We must emphasise that the nature of the convergence with the clearly defined optimal intersection regions is such that all three parameters can be optimally determined. We are not left with a set of alternate fits, as is often the case with a pole-residue plot or with some methods of analysing simulations.

4. Results for Several Systems

In this section results for some of the standard systems of critical phenomena will be summarised. Series analysis for systems with exact solutions is important for benchmaking, see [2] for a review of the 2D Ising and Baxter-Wu cases, but emphasis will be placed below on results for systems that do not have exact solutions.

4.1 Ising models

The three dimensional Ising model has been the main "realistic" test for new methods of series analysis and simulations. There is not space here to quote all the results obtained from series and simulations for this model so a biased selection is given. Early results for the simple-cubic critical point include $K_c = k_B T_c/J = 0.221654 \pm 0.000005$ from the Monte Carlo renomalization group simulation of Pawley *et al.* [13], and 0.221655 ± 0.000005 from a graphical Roskies-transform analysis by Adler [14]. More recently Liu and Fisher [15], (using inhomogeneous differential approximants), found a central threshold of 0.221630, and Ferrenberg and Landau [16] found 0.2216595 ± 0.0000026 from simulations. This latest value has a precision that is considerably beyond that obtainable with current lengths of series, but overlaps with Adler's result.

In higher dimensions, the series are more than competitive with results obtained via simulations. Series to 15th order for the pure system were obtained as a by-product of the generation of new series for the Ising model in a random field (Gofman *et al.*[17]), and were analysed in [18,19] in parallel with simulations. The simulations in [18,19] were of a slightly unsusal kind, in that critical temperatures were determined from relaxation of the magnetization.

In five dimensions $K_c = 0.113935 \pm 0.000015$ was found from these series (the best fit was obtained for $\gamma = 1.0005 \pm 0.005$ and $\Delta_1 = 0.4 \pm 0.1$, the exact values being 1 and 1/2 respectively) to be compared with 0.11391 ± 0.00001 from the simulations [18]. Older 11 term series analysed by Guttmann [20] based on imposing $\gamma = 1.0$, gave 0.11392 and $\Delta_1 = 0.5 \pm 0.05$ compatible with the simulations of [18]. This led us to be a little concerned about the series, since longer series should give better results, but the recent simulations by Rickwardt *et al.* [21] give 0.113929 ± 0.000045 confirming the series value of [18]. In six dimensions our series gave 0.092295 ± 0.000003 whereas simulations suggest 0.09230 ± 0.00005 and in seven dimensions the results are 0.077706 ± 0.000002 and 0.07772 ± 0.00003 respectively [19]. The simulations of [18,19], which when made two years ago were the largest ever simulated in those dimensions, used a parallel machine with 32 Intel i860 processors whereas the series were obtained as a byproduct of another study. The longer series for the the four dimensional Ising case requires analysis that allows for logarithmic corrections which is being undertaken at the moment.

4.2 Heisenberg models

Prior to 1993 published results for the critical temperature from series expansions up to 12th order for the three-dimensional classical Heisenberg and XY models did not agree very well with high-precision Monte Carlo estimates. The series result was $K_c = 0.6924(2)$ [22]. The simulation estimates were 0.6929 from the Metropolis method [23], 0.6930(1) from a cluster method [24] and 0.693035(37) from a single cluster multiple histogram method [25]. In order to clarify this discrepancy Adler *et al.* [26] analyzed extended high-temperature series expansions of the susceptibility, the second correlation moment, and the second field derivative of the susceptibility, which were derived by Lüscher and Weisz [27] and Butera *et al.* [28] for general $O(n)$ vector spin models on D-dimensional hypercubic lattices up to 14th order in $K \equiv J/k_BT$. We found $K_c = 0.6929 \pm 0.0001$, in good agreement with the simulations, although substantially less precise. Exponents were in good agreement with standard field theory exponent estimates.

4.3 XY models

In [26] we also reanalysed the XY model series to find $K_c = 0.45414 \pm 0.00007$. More recently, Butera, Comi and Guttmann [29] have extended the XY series to 17 terms and found exponent values in good agreement with the field theory and simulation ones. The simulation temperatures are again more precise and we quote 0.45421(8) [30], and 0.4542(1) [31].

The first-ever experimental observation of a non-analytic correction-to-scaling was made in superfluid ^{4}He, which is in the same universality class as the XY model. From the shorter 14-term series we were unable to observe any clear intesection indicating the non-analytic correction, but a reanalysis of the series of [29] with M2 clearly shows that near $K_c = 04541$ there is an intersection near $\Delta_1 = 0.5$, in good agreement with the renormalization group and experimental result.

4.3 Percolation

Series expansions for general moments of the bond percolation cluster size distribution on hypercubic lattices to 15th order in the concentration were obtained by Adler *et al* [32] in 1990. This was one more term than the previously published series for the mean cluster size in 3D and four terms more for higher moments and higher dimensions. Critical exponents, amplitude ratios and thresholds were calculated from these and other series by a variety of independent analysis techniques, and compared with other results.

In Figure 1 an example of the M2 analysis is shown for five-dimensional bond percolation. Here optimal convergence occurs for $p_c = 0.11819 \pm 0.00004$. This theshold was deduced from several measurements on different moments with both M1 and M2 and remains the most accurate five-dimensional threshold to the best of my knowledge. The exponent of the mean-cluster size can be seen to be $\gamma = 1.185 \pm 0.005$ and the correction exponent 0.55 ± 0.15 in excellent agreement with ϵ-expansion values [33] of 1.18 and 0.45, respectively. Older simulation results [34] gave $\gamma = 1.3 \pm 0.1$ Accurate exponents and thresholds were found in dimensions 3-9, and in the most interesting case of three dimensions the series gave $p_c = 0.2488 \pm 0.0002$, which was in contradiction with the then-accepted 0.2493 [35]. This caused us a delay in submission while we tried to understand the series and simulation values could be so far off, but we later heard that Ziff and Stell found 0.248810 ± 0.00005. The exponents agree with ϵ-expansion values in the high dimensions and Ziff and Stell in 3D where $\gamma = 1.805 \pm 0.02$ and $\nu = 0.872 \pm 0.070$.

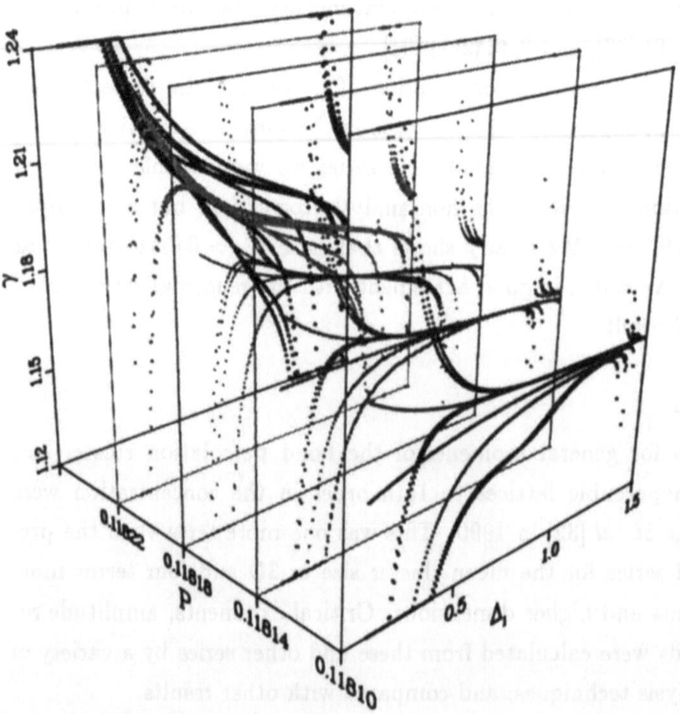

Figure 1. Plot of approximants to the dominant exponent γ as a fuction of correction-to-scaling exponent Δ_1, from the M2 analysis at five trial thresholds for the susceptibilty of five dimensional percolation.

5. Conclusions

There is a consistent pattern in these models of more precision from recent simulations in two or three dimensions, with the simulation estimates falling well within the series error bounds. There is as yet no case, even amongst more complex systems, where a series analysis that allowed for corrections-to-scaling failed to include later precise simulation results. In higher dimensions, the situation is reversed, with higher precision being found for the series. In every case the series results are obtained with orders of magnitude less computer resources, (but not always less manpower.) It is comforting that as things stand at present there remain no serious discrepancies between series and simulations in the simple models discussed herein.

Looking to the future one must think about what extensions are possible in both approaches. Extensions to sample size or series length do not necessarily imply automatic improvement as both series and simulations are also dependent on anlysis method and the quantities studied. For example: all n-vector simulations except [18,19] used sophisticated histogram analysis techniques which give higher quality results from a given size. Low temperature series for the Ising model are much longer than the high temperature ones, but give far worse convergence in all cases. However for a given analysis method larger simulations or longer series usually give better results, as opposed to the case with some renormanlization group methods.

The series discussed above (all high temperature ones) range from 15 thru to 21 terms. A significant extension to the series is unlikely in the near future but 3-5 more terms would be feasible with current computers and algorithms. A renalysis of some recently extended low temperature series for the 3D Ising is currently being carried out by the author. The simulations reported here range from $L \leq 96$ in [16] for the 3D Ising, and $L \leq 40$ for the 3D Heisenberg [25], to $L \leq 17$ [21] and $L \leq 45$ [18]. Further increments are possible to the sizes (the simulations of [16] are currently being extended). The question concerning extension is more one of whether we are now precise enough in terms of testing conjectures and theories. The answer to this is definitely no in two cases; one is the 5D Ising, where some finite size scaling matters remain unchecked the other the 3D Ising where the conjecture of Rosengren [36] $K_c = 0.22165863$ remains consistent with [14] and [16]. Thus further studies even of some simple models are desirable.

The real excitement for many of us will be in the extension and application of both methods for the many questions of interest in more complex models that include quantum effects and disorder.

Acknowledgements: I thank the BSF and GIF for support during many of the calculations described above. Discussions with all my co-authors of the results reviewed here, and many others active in series and simulations were very helpful.

References

1. J. Adler, Computers in Physics, **8**, 287 (1994).
2. J. Adler, for "Annual Reviews of Computational Physics", edited by D. Stauffer, (1995).

3. J. M. Yeomans, *Statistical Mechanics of Phase Transitions*, Clarendon Press (1992).
4. H. E. Stanley, *Introduction to Phase Transitions and Critical Phenomena*, Oxford University Press (1971). See especially Chapter 8.
5. A. J. Guttmann in *Phase Transitions and Critical Phenomena*, Vol. 13, C. Domb and J. L. Lebowitz eds., Academic Press, New York (1989).
6. Descriptions of early series expansions for many systems can be found in *Phase Transitions and Critical Phenomena* Vol. 2, C. Domb and M. S. Green eds., Academic Press, New York (1972), *ibid.*,Vol. 3 (1974).
7. J. Adler, I. Chang and S. Shapira, *Int. J. Mod. Phys.* C **4**, 1007 (1993).
8. J. Adler, M. Moshe and V. Privman, Chapter 17 of *Percolation Structures and Processes*, Annals of the Israel Physical Society, Vol. 5, G. Deutscher, R. Zallen, and J. Adler eds., Adam Hilger, London (1983).
9. G. A. Baker, *Quantitative Theory of Critical Phenomena*, Academic Press (1990). This reference is the most complete text for all aspects of series expansions.
10. J. Adler, Y. Meir, A. Aharony, A. B. Harris, and L. Klein, J. Stat. Phys. **58**, 511 (1990).
11. M. E. Fisher and R. R. P. Singh, *Disorder in Physical Systems*, G. Grimmett and D. J. A. Welsh eds., Oxford University Press (1990) in honor of J. M. Hammersley.
12. M. E. Fisher and J.-H. Chen in *Phase Transitions: Cargése 1980*, edited by M. Levy, J. C. Le Guillou and J. Zinn-Justin, (Plenum, New York, 1982), as well as other chapters of this volume.
13. G. S. Pawley, R. H. Swendsen, D. J. Wallace and K. G. Wilson, Phys. Rev. B **29**, 4030 (1984).
14. J. Adler, J. Phys. A **16**, 3835 (1983).
15. A. J. Liu and M. E. Fisher, Physica A **156**, 35 (1989).
16. A. M. Ferrenberg and D. P. Landau, Phys. Rev. B **44**, 5081 (1991).
17. M. Gofman, J. Adler, A. Aharony, A. B. Harris and M. Schwartz, Phys. Rev. Lett., **71**, 1569 (1993).
18. C. Munkel, D. Heermann, J. Adler, M. Gofman and D. Stauffer, Physica A **193**, 540 (1993).
19. M. Gofman, J. Adler, A. Aharony, A. B. Harris and D. Stauffer, J. Stat. Phys, **71**, 1219 (1993).
20. A. J. Guttmann, J. Phys. A., **14**, 22, (1981).

21. Ch. Rickwardt, P. Nielaba and K. Binder, Ann. Physik (Leipzig) **3**, 483 (1994).
22. S. Mckenzie, C. Domb, and D. L. Hunter, J. Phys. A. **15**, 3899 (1982).
23. P. Peczak, A. M. Ferrenberg, and D. P. Landau, Phys. Rev. **B43**, 6087 (1991).
24. C. Holm and W. Janke, Phys. Lett. **A173**, 8 (1993), and preprint FUB-HEP 19/92, HLRZ 77/92, Berlin/Jülich (1992).
25. K. Chen, A.M. Ferrenberg, and D.P. Landau, Phys. Rev. B **48**, 3249 (1993).
26. Adler, C. Holm and W. Janke, Physica A **201**, 581 (1993).
27. M. Lüscher and P. Weisz, Nucl. Phys. **B300**, 325 (1988).
28. P. Butera, M. Comi, and G. Marchesini, Phys. Rev. **B41**, 11494 (1990).
29. P. Butera, M. Comi, and A. J. Guttmann, preprint.
30. M. Hasenbusch and S. Meyer, Phys. Lett. **B241**, 238 (1990).
31. W. Janke, Phys. Lett. **A148**, 306 (1990).
32. J. Adler, Y. Meir, A. Aharony and A. B. Harris, Phys. Rev. B, **41**, 9183 (1990).
33. J. S. Reeve, A. J. Guttmann, *Phys. Rev.* B **26**, 3923 (1982); O. F. de Alcantara Bonfim, J. E. Kirkham, and A. J. McKane, *J. Phys.* A **13**, L247 (1980); *ibid.* **14**, 2391 (1981).
34. S. Kirkpatrick, Phys. Rev. Lett., **36**, 69 (1976).
35. D. Stauffer and J. G. Zabolitsky, J. Phys. A **19**,3705 (1987); S. Wilke, Phys. Lett. **96A**, 344 (1983).
36. A. Rosengren, J. Phys. A **19**, 1709 (1986).

Applications of Finite-Size-Scaling Techniques to the Simulation of Critical Fluids

N.B. Wilding

Institut für Physik, Universität Mainz,
Staudinger Weg 7, D-55099 Mainz, Germany

Abstract. A finite-size scaling theory is described that takes account of the lack of symmetry between the coexisting phases of fluids. This broken symmetry is manifest in the so-called 'field mixing' phenomenon which is a central feature of the non-universal critical behaviour of fluids. It is shown that the presence of field mixing leads to an alteration to the limiting form of the critical energy distribution and to a finite-size correction to the critical order parameter (particle density) distribution. As a result, finite-size shifts occur in the critical particle and energy densities. The theoretical predictions are tested with an extensive Monte-Carlo study of the critical density and energy fluctuations of the 3D Lennard-Jones fluid.

1 Introduction

In the absence of exact analytical solutions, computer simulations provides an invaluable route to obtaining the phase coexistence properties of model systems, knowledge of which is important for elucidating physical behaviour and testing theoretical predictions. Over the years, great progress has been made in the development of simulation techniques for studying phase behaviour, and accurate phase properties are now available for a whole host of simple model systems. Notwithstanding these success, however, one important aspect of phase behaviour in model systems continues to pose a challenge to computer simulation, namely the critical point regime. In the vicinity of a critical point, the correlation length ξ grows extremely large and may exceed the linear size L of the simulated system. When this occurs, the singularities and discontinuities that characterise critical phenomena in the thermodynamic limit are shifted and smeared out. Unless care is exercised, such finite-size effects can lead to serious errors in computer simulation estimates of critical point parameters.

To deal with these problems finite-size-scaling (FSS) techniques have been developed. FSS methods enable one to extract accurate estimates of infinite-volume quantities from simulations of finite-size. Of the many previous FSS simulation studies that have been performed [1, 2], most have focussed on critical phenomena in lattice-based magnetic spin systems such as the Ising [3, 4], ϕ^4 [5], XY [6] and Heisenberg models [7, 8]. Among the specific approaches employed to study such systems, use of the order parameter (magnetisation) distribution function has proved itself one of the most powerful. The FSS properties of the order parameter distribution are now routinely employed in studies of magnetic systems, facilitating both the accurate location of the critical point and the detailed elucidation of its character [2].

Only comparatively recently has attention turned to the task of applying FSS techniques to the simulation of critical fluids. Work to date has concentrated on attempting to carry over to fluids the order parameter distribution techniques developed in the magnetic context [9–14] In the process, however, it has been found necessary to generalise the FSS

Springer Proceedings in Physics, Vol. 80
Computer Simulation Studies in Condensed-Matter Physics VIII
Editors: D.P. Landau · K.K. Mon · H.-B. Schüttler

equations to take account of the absence in fluids of the energetic ('particle-hole') symmetry that prevails in magnetic systems such as the Ising model. Within the renormalisation group framework, this broken symmetry is expressed (as described below) through the mixed character of the relevant scaling fields that control the critical behaviour. Although such 'field mixing' is believed to have no bearing on the universal critical behaviour of fluids, (which for system with short-ranged interactions correspond to the Ising universality class), it has long been recognised that it leads to certain non-universal effects. The most celebrated of these is a weak critical singularity in the coexistence diameter, the existence of which constitutes a failure for the law of rectilinear diameter [15].

In this paper we investigate the consequences of field mixing for the FSS properties of near critical fluids. We begin in section 2 by providing a brief introduction to the ideas of finite-size-scaling and demonstrate how one incorporates field mixing effects. In section 3 we use Monte Carlo studies of the 3D Ising model to predict the consequences of field mixing for the critical order-parameter and energy distributions of fluids. Extensive Monte Carlo simulation studies of the 3D Lennard-Jones fluid are then used to test the theoretical predictions in detail. It is shown that application of mixed-field FSS techniques permits a very accurate determination of the critical point parameters of model fluids.

2 Background

As its basic tenant, the FSS hypothesis holds that for sufficiently large ξ and L, the coarse-grained properties of a given near-critical system are universal and depend (up to certain non-universal factors) only on specific combinations of L and the relevant scaling fields that measure deviations from criticality [16]. For single component systems such as simple fluids and magnets, there are two relevant scaling fields which we denote τ and h. In models of the Ising symmetry, τ is simply the reduced temperature, while h is the reduced magnetic field. Conjugate to each of the scaling fields are scaling operators $\mathcal{M}$ and $\mathcal{E}$. In the special case of the Ising model, one finds $\mathcal{M} \rightarrow m$ (the magnetisation) and $\mathcal{E} \rightarrow u$ (the energy density).

Finite-size scaling arguments [12], with a solid renormalisation group basis [17], predict that the joint distribution $p_L(\mathcal{M}, \mathcal{E})$ takes the form

$$p_L(\mathcal{M}, \mathcal{E}) \simeq \Lambda_{\mathcal{M}}^{+} \Lambda_{\mathcal{E}}^{+} \tilde{p}_{\mathcal{M},\mathcal{E}}(\Lambda_{\mathcal{M}}^{+} \delta\mathcal{M}, \Lambda_{\mathcal{E}}^{+} \delta\mathcal{E}, \Lambda_{\mathcal{M}} h, \Lambda_{\mathcal{E}} \tau) \tag{2.1a}$$

where

$$\Lambda_{\mathcal{E}} = a_{\mathcal{E}} L^{1/\nu} \qquad \Lambda_{\mathcal{M}} = a_{\mathcal{M}} L^{d-\beta/\nu} \qquad \Lambda_{\mathcal{M}} \Lambda_{\mathcal{M}}^{+} = \Lambda_{\mathcal{E}} \Lambda_{\mathcal{E}}^{+} = L^d \tag{2.1b}$$

and

$$\delta\mathcal{M} \equiv \mathcal{M} - <\mathcal{M}>_c \qquad \delta\mathcal{E} \equiv \mathcal{E} - <\mathcal{E}>_c \tag{2.1c}$$

The subscripts c in equations 2.1c signify that the averages are to be taken at criticality. Given appropriate choices for the non-universal scale factors $a_{\mathcal{M}}$ and $a_{\mathcal{E}}$ (equation 2.1b), the function $\tilde{p}^{\star}_{\mathcal{M},\mathcal{E}}$ is expected to be universal.

Precisely at criticality, the scaling fields τ and h vanish by definition, implying that

$$p_L(\mathcal{M}, \mathcal{E}) \simeq \Lambda_{\mathcal{M}}^{+} \Lambda_{\mathcal{E}}^{+} \tilde{p}^{\star}_{\mathcal{M},\mathcal{E}}(\Lambda_{\mathcal{M}}^{+} \delta\mathcal{M}, \Lambda_{\mathcal{E}}^{+} \delta\mathcal{E}), \tag{2.2}$$

where $\tilde{p}^{\star}_{\mathcal{M},\mathcal{E}}(x, y) = \tilde{p}^{\star}_{\mathcal{M},\mathcal{E}}(x, y, 0, 0)$ is a function describing the universal and statistically scale invariant operator fluctuations characteristic of the critical point. It follows that $\tilde{p}^{\star}_{\mathcal{M},\mathcal{E}}(x, y)$ constitutes a ***hallmark*** of a universality class, a fact that can be exploited to obtain accurate estimates of the critical point parameters of fluid systems.

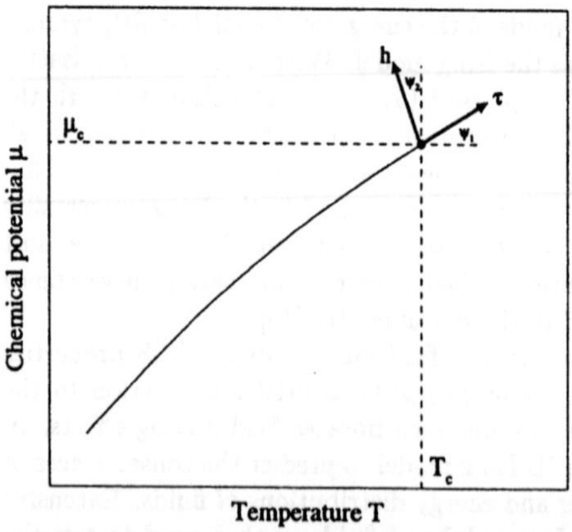

Figure 1: Schematic representation of the liquid-vapour coexistence curve showing the directions of the relevant scaling scaling fields. The angles ψ_1 and ψ_2 are related to the field-mixing parameters s and r (equation 2.3) by $r = -\tan\psi_1$ and $s = \tan\psi_2$.

Although the generality of the FSS expression, equation 2.1a, encompasses all members of a given universality class, irrespective of whether or not they possess the special Ising symmetry, it has long been appreciated [18] that the *form* of the fluid scaling fields differ from those of the Ising model. Specifically, the lack of symmetry between the coexisting phases of fluids leads to 'mixed' scaling fields comprising *linear-combinations* of the reduced (inverse) temperature and applied (chemical potential) field:

$$\tau = T - T_c + s(\mu - \mu_c) \qquad h = \mu - \mu_c + r(T - T_c) \tag{2.3}$$

where the parameters s and r are system-specific quantities controlling the degree of field mixing and the subscript c signifies critical values. In particular r is identifiable as the limiting critical gradient of the coexistence curve in the space of μ and T. The role of s is somewhat less tangible; it controls the degree to which the chemical potential features in the thermal scaling field, manifest in the widely observed critical singularity of the coexistence curve diameter of fluids [15]. The directions of the fluid scaling fields are indicated schematically in the phase diagram of figure 1.

As a result of the mixed character of the fluid scaling fields, the conjugate scaling operators are also found to comprise linear combinations [12] of the order parameter (particle density ρ) and the energy density u. Specifically, one finds

$$\mathcal{M} = \frac{1}{1-sr}\left[\rho - su\right] \qquad \mathcal{E} = \frac{1}{1-sr}\left[u - r\rho\right]. \tag{2.4}$$

The forms of the scaling fields and scaling operators for systems with and without field mixing are summarised in table 1.

The operators $\mathcal{M}$ and $\mathcal{E}$ are the respective scaling counterparts of the Ising magnetisation and energy. Typically, however, in fluid simulations we are more concerned with obtaining the critical density and energy density. Owing to field mixing, these quantities are not expected to exhibit the same FSS properties as the Ising magnetisation or energy, and thus it is necessary to examine their behaviour more closely.

Table 1: The forms of the relevant scaling fields and scaling operators for the Ising magnet and for pure fluids.

Ising model	Fluids
$\tau = T - T_c$	$\tau = T - T_c + s(\mu - \mu_c)$
$h = H - H_c$	$h = \mu - \mu_c + r(T - T_c)$
$\mathcal{E} = u$	$\mathcal{E} = \frac{1}{1-sr}[u - r\rho]$
$\mathcal{M} = m$	$\mathcal{M} = \frac{1}{1-sr}[\rho - su]$

3 Monte-Carlo studies

3.1 The universal finite-size spectrum of $p_L(\rho)$ and $p_L(u)$

In what follows we shall employ Monte Carlo simulations to investigate the consequences of field mixing for the order parameter and energy distributions of near-critical fluids. To do so, however, we first require an independent estimate of the fixed point function $\tilde{p}^{\star}_{\mathcal{M},\mathcal{E}}(x,y)$ appropriate to the Ising universality class. In practice, this function is most readily obtained by considering the prototype member of the Ising class, namely the 3D Ising model itself. Owing to its lack of field mixing, the scaling operators of the Ising model are accessible simply as the magnetisation and energy density. Moreover, the availability of highly accurate estimates for the Ising model critical temperature, circumvents the need to perform a time consuming search for the critical point.

Using a vectorised algorithm on a Cray YMP supercomputer, we have performed high precision Monte Carlo simulation measurements of the joint magnetisation and energy density distribution for the 3D Ising model on a periodic lattice of side $L = 32$. The measurements were performed at the estimated (reduced) critical coupling $K_c^I = 0.2216595(26)$, as obtained in a previous high precision Monte Carlo study [4]. The resulting form of $\tilde{p}^{\star}_{\mathcal{M},\mathcal{E}}$, normalised to unit integrated weight and scaled to unit variance along both axes, is shown in figure 2. The associated ordering operator distribution [19] $\tilde{p}^{\star}_{\mathcal{M}}(x) = \int \tilde{p}^{\star}_{\mathcal{M},\mathcal{E}}(x,y)dy$, and energy operator distribution $\tilde{p}^{\star}_{\mathcal{E}}(y) = \int \tilde{p}^{\star}_{\mathcal{M},\mathcal{E}}(x,y)dx$ are shown in figure 3. One observes that while the form of $\tilde{p}^{\star}_{\mathcal{M}}(x)$ is doubly peaked and symmetric, that of $\tilde{p}^{\star}_{\mathcal{E}}(y)$ is singly peaked and asymmetric.

Knowledge of the form of $\tilde{p}^{\star}_{\mathcal{M},\mathcal{E}}(x,y)$ can be used to obtain the universal finite-size spectrum of the density and energy fluctuations of critical fluids. To facilitate this, it is expedient to reexpress ρ and u in terms of the scaling operators. Appealing to equation 2.4, one finds

$$u = \mathcal{E} - r\mathcal{M} \qquad \rho = \mathcal{M} - s\mathcal{E}, \tag{3.1}$$

so that the critical density and energy density distributions are

$$p_L(u) = p_L(\mathcal{E} - r\mathcal{M}) \qquad p_L(\rho) = p_L(\mathcal{M} - s\mathcal{E}). \tag{3.2}$$

Now the structure of the scaling form 2.1a shows that the typical size of the fluctuations in the energy-like operator will vary with system size like $\delta\mathcal{E} \sim L^{-(1-\alpha)/\nu}$, while the typical size of the fluctuations in the ordering operator vary like $\delta\mathcal{M} \sim L^{-\beta/\nu}$. It follows that for a given L, the shape of the energy and density distributions can be identified with the distribution of the variable

$$X_\Theta = a_{\mathcal{M}}^{-1}\delta\mathcal{M}\cos\Theta + a_{\mathcal{E}}^{-1}\delta\mathcal{E}\sin\Theta, \tag{3.3}$$

with

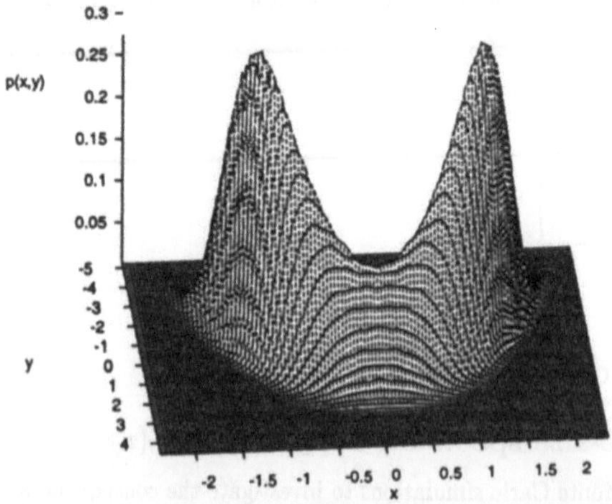

Figure 2: Estimates of the fixed point form of the joint scaling operator distribution $\tilde{p}^{\star}_{\mathcal{M},\mathcal{E}}$ appropriate to the 3D Ising universality class, obtained as the joint magnetisation and energy density distribution of the $L = 32$ periodic 3D Ising model at the estimated critical coupling $K = 0.2216595$. The data is expressed in terms of the scaling variables $x = a_{\mathcal{M}}^{-1}L^{\beta/\nu}(\mathcal{M} - \mathcal{M}_c)$ and $y = a_{\mathcal{E}}^{-1}L^{(1-\alpha)/\nu}(\mathcal{E} - \mathcal{E}_c)$, with the scale factors $a_{\mathcal{E}}^{-1}$ and $a_{\mathcal{M}}^{-1}$ assigned so that the distribution has unit variance along both axes.

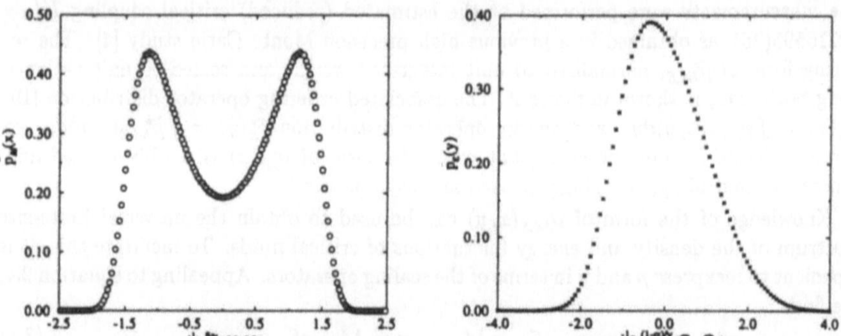

Figure 3: (a) The ordering operator distribution $\tilde{p}^{\star}_{\mathcal{M}}(x)$ appropriate to the 3D Ising universality class, obtained as the magnetisation distribution of the $L = 32$ 3D periodic Ising model at the assigned value of the critical coupling $K_c^I = 0.2216595$, and expressed in terms of the scaling variable $x = a_{\mathcal{M}}^{-1}L^{\beta/\nu}(\mathcal{M} - \mathcal{M}_c)$. (b) The corresponding energy operator distribution $\tilde{p}^{\star}_{\mathcal{E}}(y)$, expressed in terms of the scaling variable $y = a_{\mathcal{E}}^{-1}L^{(1-1/\nu)}(\mathcal{E} - \mathcal{E}_c)$. In both cases, the value of the non-universal scale factors $a_{\mathcal{E}}^{-1}$ and $a_{\mathcal{M}}^{-1}$ were chosen so that the distributions have unit variance. Statistical errors do not exceed the symbol sizes.

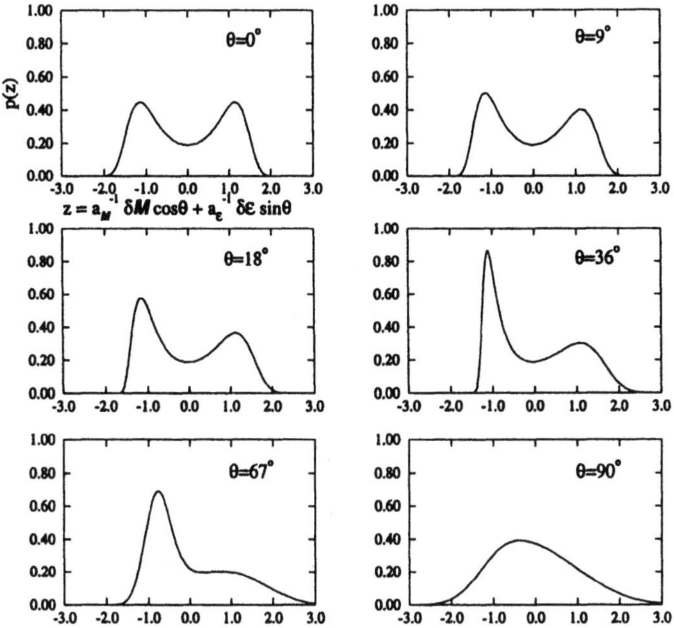

Figure 4: Selections from the universal finite-size spectrum of critical density and energy density distributions of fluids. The distributions were obtained according to the procedure described in the text. Following convention, the values of the non-universal scale factors $a_{\mathcal{E}}^{-1}$ and $a_{\mathcal{M}}^{-1}$ have been chosen to ensure that the distributions have unit variance.

$$\tan\Theta_u = \frac{a_{\mathcal{E}}}{r a_{\mathcal{M}}} L^{-(1-\alpha-\beta)/\nu} \quad , \quad \tan\Theta_\rho = \frac{s a_{\mathcal{E}}}{a_{\mathcal{M}}} L^{-(1-\alpha-\beta)/\nu} \tag{3.4}$$

where the subscripts u and ρ signify that the value of Θ corresponds to the energy density and density distributions respectively.

The distributions $p(X_\Theta)$ constitute a one-parameter class (spectrum) of *universal* functions describing the density and energy distributions of fluids at finite L. Heueristically, this L dependence arises from the different relative strengths of the critical fluctuations in $\mathcal{M}$ and $\mathcal{E}$. Since $1-\alpha > \beta$, the critical fluctuations in $\mathcal{E}$ are weaker than those in $\mathcal{M}$, causing the distribution $p_L(\mathcal{E})$ to converge to its average value more rapidly with increasing L, than $p_L(\mathcal{M})$. As a result, measurements of the distributions of ρ and u, (each of which represent linear combinations of $\mathcal{M}$ and $\mathcal{E}$), yield L-dependent functional forms.

Geometrically, the parameter Θ can be interpreted as defining a direction OX_Θ in the basal plane formed by the Ox and Oy axes of figure 2, making an angle Θ with the Ox axis. The form of $p(X_\Theta)$ is then obtainable by projecting $\tilde{p}^*_{\mathcal{M},\mathcal{E}}(x,y)$, onto the vertical plane which includes the line OX_Θ. A representative selection of such projections is shown in figure 4. For $\Theta = 0°$ one obtains simply the ordering operator distribution $\tilde{p}^*_{\mathcal{M}}(x)$, while for $\Theta = 90°$ the from is that of the energy-like operator $\tilde{p}^*_{\mathcal{E}}(y)$ distribution. Intermediate between these values a range of behaviour is obtained, representing the finite L forms of $p_L(\rho)$ and $p_L(u)$.

Asymptotically (i.e. as $L \to \infty$), equation 3.4 implies that both Θ_u and Θ_ρ approach zero [20] so that in this limit

$$p_L(u) = p_L(-r\mathcal{M}) \simeq a_{\mathcal{M}}{}^{-1} r L^{\beta/\nu} \tilde{p}^{\star}_{\mathcal{M}}(-a_{\mathcal{M}}{}^{-1} r L^{\beta/\nu} \delta\mathcal{M}) \tag{3.5}$$

$$p_L(\rho) = p_L(\mathcal{M}) \simeq a_{\mathcal{M}}{}^{-1} L^{\beta/\nu} \tilde{p}^{\star}_{\mathcal{M}}(a_{\mathcal{M}}{}^{-1} L^{\beta/\nu} \delta\mathcal{M}) \tag{3.6}$$

It follows that for *any* finite s and r, the limiting critical point forms of $p_L(\rho)$ and $p_L(u)$ both match the critical ordering operator distribution $\tilde{p}^{\star}_{\mathcal{M}}(x)$. This of course simply reflects the fact that for sufficiently large L the critical fluctuations in $\mathcal{E}$ are negligible on the scale of those in $\mathcal{M}$. We note, however, that such a state of affairs differs radically from the case of the critical Ising model where, owing to the absence of field mixing, $\lim_{L\to\infty} p_L(u) = \tilde{p}^{\star}_{\mathcal{E}}(y)$. This finding has important ramifications for calculations of the specific heat. In models of the Ising symmetry the specific heat at criticality (as calculated from the variance of the energy distribution), scales with system size like

$$C_v = L^d(\langle u^2\rangle - \langle u\rangle^2)/k_B T^2 \propto L^{-\alpha/\nu} \tag{3.7}$$

By contrast for fluids, the alteration to the limiting form of the critical energy distribution implies that the specific heat scales asymptotically like the Ising *susceptibility* (fluid compressibility)

$$C_v = L^d(\langle u^2\rangle - \langle u\rangle^2)/k_B T^2 \propto L^{-\gamma/\nu} \tag{3.8}$$

It follows that if one wishes to measure the exponent α from the FSS behaviour of critical fluids, it is necessary to study the behaviour of the energy *operator* distribution $p_L(\mathcal{E})$ and not the energy itself.

Finally in this subsection, we point out that precise knowledge of the location of the critical point, does *not* imply the possibility of directly measuring the infinite-volume density and energy density. To appreciate this, recall that

$$\langle u\rangle_c = \langle\mathcal{E}\rangle_c - r\langle\mathcal{M}\rangle_c \qquad \langle\rho\rangle_c = \langle\mathcal{M}\rangle_c - s\langle\mathcal{E}\rangle_c \tag{3.9}$$

Now, while symmetry considerations dictate that the value of $\langle\mathcal{M}\rangle_c = \int p_L(\mathcal{M})d\mathcal{M}$ is independent of L, no such symmetry condition pertains to $p_L(\mathcal{E})$, whose average value $\langle\mathcal{E}\rangle_c = \int d\mathcal{E} p_L(\mathcal{E})$ at criticality is expected to vary with system size like

$$\langle\mathcal{E}\rangle_c(L) - \langle\mathcal{E}\rangle_c(\infty) \sim L^{-(1-\alpha)/\nu} \tag{3.10}$$

It follows that in order to extract infinite volume estimates of ρ_c and u_c from simulations at the critical point, it is necessary to extrapolate data from a number of different system sizes to the thermodynamic limit [21].

3.2 Application to the Lennard-Jones fluid

Monte-Carlo simulations were performed on the 3D Lennard-Jones fluid using a Metropolis algorithm within the grand canonical ensemble [22]. The algorithm employed is similar in form to that described by Adams [23, 24], but differs in the respect that only particle transfer (insertion and deletion) steps were implemented, leaving particle moves to be performed implicitly as a result of repeated transfers. Physically this choice is motivated by the need to direct the computational effort at the density fluctuations, which are the bottleneck for phase space evolution at criticality.

As is common practice in simulations of systems whose interparticle potential decays rapidly with particle separation, the Lennard-Jones potential was truncated in order to reduce the computational effort. In accordance with most previous studies of the LJ system, the cutoff radius was chosen to be $r_c = 2.5\sigma$, and the potential was left unshifted. It should be noted, however, that the choice of cutoff can has quite marked effects on the critical point parameters, a point emphasised by Smit [25]. In the course of the simulations, a total of 5 system sizes were studies corresponding to $L = mr_c$ with $m = 3, 4, 5, 6, 7$. For the observables to be recorded, the particle density ρ and the energy density u were chosen. The joint distribution $p_L(\rho, u)$ was accumulated in the form of a histogram.

A recent previous simulation study of the LJ fluid, (using $r_c = 2.5\sigma$), places the reduced critical temperature at $T^* = 1.176(8)$ [26]. Using this estimate, we attempted to locate the liquid-vapour coexistence curve by performing a series of very short runs for the $m = 4$ system size, in which the effective chemical potential μ^* was tuned until the density distribution exhibited a double peaked structure. Having obtained, in this manner, an approximate value of the coexistence chemical potential, a longer run comprising 2×10^7 MCS was performed to accumulate better statistics. In order to explore the coexistence curve in the neighbourhood of this simulation temperature, the histogram reweighting technique of Ferrenberg and Swendsen [27] was employed. This technique allows histograms of observables accumulated at one set of model parameters, to be reweighted to yield estimates for histograms appropriate to another set of parameters. The histogram reweighting method is especially profitable close to the critical point where, owing to the large critical fluctuations, a single simulation affords reliable extrapolations over the entire critical region.

To facilitate a precise identification of the coexistence chemical potential, we adopted the criterion that the ordering operator distribution $p_L(\mathcal{M}) = \int d\mathcal{E} p_L(\mathcal{M}, \mathcal{E})$ must be symmetric in $\mathcal{M} - \langle\mathcal{M}\rangle$. This criterion is the counterpart of the coexistence symmetry condition for the Ising model magnetisation distribution. By simultaneously tuning μ and s in the reweighting of the joint distribution $p_L(\rho, u)$, estimates for the coexistence chemical potential and the value of the field mixing parameter s that satisfy this symmetry condition, were readily obtained.

To obtain a preliminary estimate of the critical point parameters, the universal matching condition for the ordering operator distribution $p_L(\mathcal{M})$ was invoked. As observed in section 2, fluid–magnet universality implies that the critical fluid ordering operator distribution $p_L(\mathcal{M})$ must match the universal fixed point function $\tilde{p}^*_{\mathcal{M}}(x)$ appropriate to the Ising universality class. Leaving aside for present the question of corrections to scaling, the apparent critical point of the fluid can thus be estimated by tuning the temperature, chemical potential and field mixing parameter s (within the reweighting scheme) such that $p_L(\mathcal{M})$ collapses onto $\tilde{p}^*_{\mathcal{M}}(x)$. The result of applying this procedure for the $m = 4$ data set is displayed in figure 5 where the data has been expressed in terms of the scaling variable $x = a_{\mathcal{M}}^{-1} L^{\beta/\nu}(\mathcal{M} - \mathcal{M}_c)$. The accord shown corresponds to a choice of the apparent critical parameters $T^*_c(L) = 1.1853(2), \mu^*_c(L) = -2.7843(3)$.

Using these estimates of the critical parameters, extensive simulations were then performed for each of the 5 system sizes $m = 3-7$ in order to facilitate a full finite-size-scaling analysis. Reweighting was again applied to the resulting histograms to effect the matching of $p_L(\mathcal{M})$ to $\tilde{p}^*_{\mathcal{M}}(x)$, thus yielding values of the apparent critical parameters. Interestingly, however, the apparent critical parameters determined in this manner were found to be *L-dependent*. The reason for this turns out to be significant contributions to the measured histograms from corrections to scaling, (not included in our scaling form 2.1a) manifest as an *L*-dependent discrepancy between the critical operator distributions and their limiting fixed point forms. In the case of the ordering operator distribution $p_L(\mathcal{M})$,

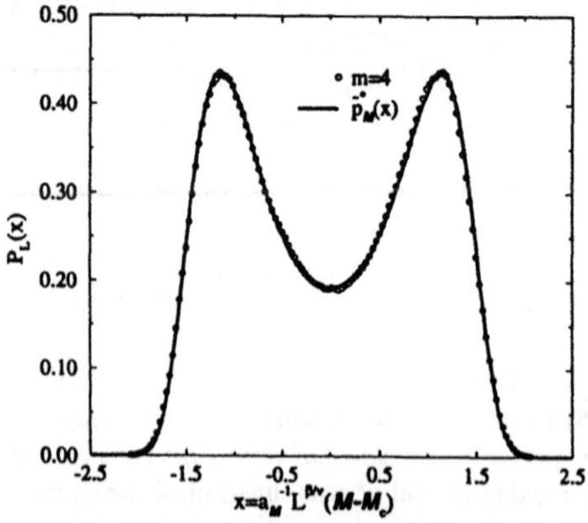

Figure 5: The measured form of the ordering operator distribution $p_L(\mathcal{M})$ for the $m = 4$ system size at the apparent critical parameters $T_c^* = 1.1853, \mu_c^* = -2.7843$. Also shown for comparison is the universal fixed point ordering operator distribution $\tilde{p}^*_{\mathcal{M}}(x)$. The data has been expressed in terms of the scaling variable $x = a_{\mathcal{M}}^{-1} L^{\beta/\nu}(\mathcal{M} - \mathcal{M}_c)$, with the value of the non-universal scale factor $a_{\mathcal{M}}^{-1}$ chosen so that the distributions have unit variance. Statistical errors do not exceed the symbol sizes.

the symmetry of the Ising problem implies that the correction to scaling function is symmetric in $\mathcal{M} - \langle\mathcal{M}\rangle$. In attempting to implement the matching to $\tilde{p}^*_{\mathcal{M}}(x)$ we therefore necessarily introduce an *additional* symmetric contribution to $p_L(\mathcal{M})$ associated with a finite value of the scaling field τ. This latter contribution has, coincidentally, a form that is very similar to that of the correction to scaling function, a result which of course makes the cancellation of contributions possible. It follows, therefore, that the magnitude of two contributions must be approximately equal.

Notwithstanding the added complications that corrections to scaling engender, it is nevertheless possible to extract accurate estimates of the infinite-volume critical parameters from the measured histograms. The key to accomplishing this is the known scaling behaviour of the corrections to scaling, which die away with increasing system size like $L^{-\theta/\nu}$, where θ is a universal correction to scaling exponent which for the 3D Ising universality class [28] takes the values $\theta \simeq 0.5$. Now, since contributions to $p_L(\mathcal{M})$ from finite values of τ, grow with system size like $\tau L^{1/\nu}$, it follows that implementation of the matching condition leads to a deviation of the apparent critical temperature $T_c^*(L)$ from the true critical temperature T_c^* which behaves like

$$T_c^*(\infty) - T_c^*(L) \propto L^{-(\theta+1)/\nu}. \tag{3.11}$$

In figure 6 we plot the apparent critical temperature $T_c^*(L)$ as a function of $L^{-(\theta+1)/\nu}$. One observes that the data is very indeed well described by a linear dependence, the least squares extrapolation of which yields the infinite-volume estimate $T_c^* = 1.1876(3)$. The associated estimate for the critical chemical potential is $\mu_c^* = -2.778(2)$. We note however, that although the coexistence value of μ^* is tightly tied to T^*, estimates of μ_c^* are not directly affected by corrections to scaling in $p_L(\mathcal{M})$, since the function $\delta p_L(\mathcal{M})/\delta\mu$ is (to leading order) antisymmetric in $\mathcal{M} - \langle\mathcal{M}\rangle$ [12].

With regard to the critical point field mixing parameters, s and r, the values of these quantities were assigned (as described in detail in reference [21]) such as to optimise the mapping of the critical operator distributions onto their limiting fixed point forms, (cf.

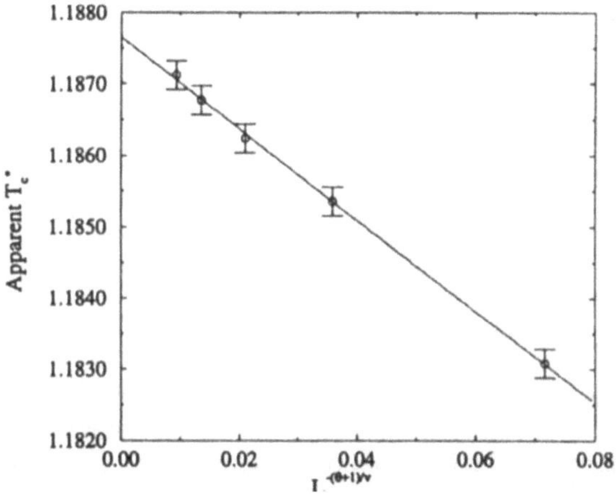

Figure 6: The apparent reduced critical temperature, (as defined by the matching condition described in the text), plotted as a function of $L^{-(\theta+1)/\nu}$, with $\theta = 0.5$ and $\nu = 0.629$ [28, 4]. The extrapolation of the least squares fit to infinite volume yields the estimate $T_c^* = 1.1876(3)$.

figure 5). The resulting estimates were, however, found to be slightly L-dependent for the smaller system sizes, an effect which we ascribe to corrections to scaling. For the two largest system sizes this L-dependence is, however, small and we find $s = -0.11(1)$, $r = -1.02(1)$.

Turning now to the critical density and energy distributions, figure 7 shows the measured forms of $p_L(\rho)$ and $p_L(u)$ at the designated critical parameters. These distributions are to varying degrees asymmetric, (a fact which as explained in the previous subsection stems from field mixing effects) and have forms that are indeed very similar to those shown in figure 4. The magnitude of this asymmetry also clearly dies away with increasing L, although the system sizes are still too small to reveal the limiting behaviour $p_L(\rho), p_L(u) \to \tilde{p}^\star_{\mathcal{M}}(x)$.

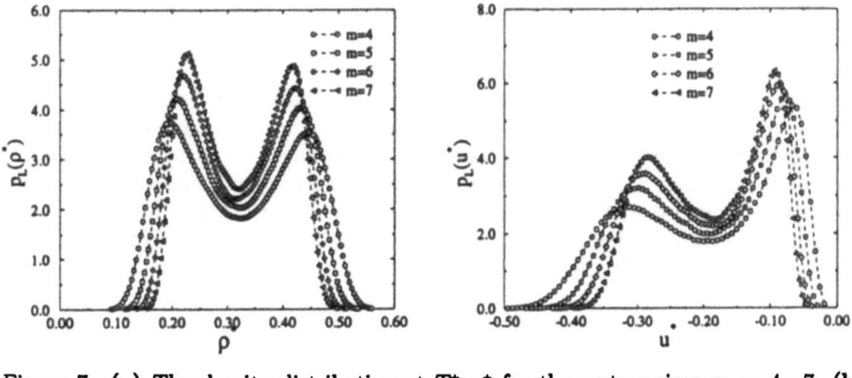

Figure 7: (a) The density distribution at T_c^*, μ_c^* for the system sizes $m = 4$—7, (b) The corresponding energy density distributions. The lines are merely guides to the eye. Statistical errors do not exceed the symbol sizes.

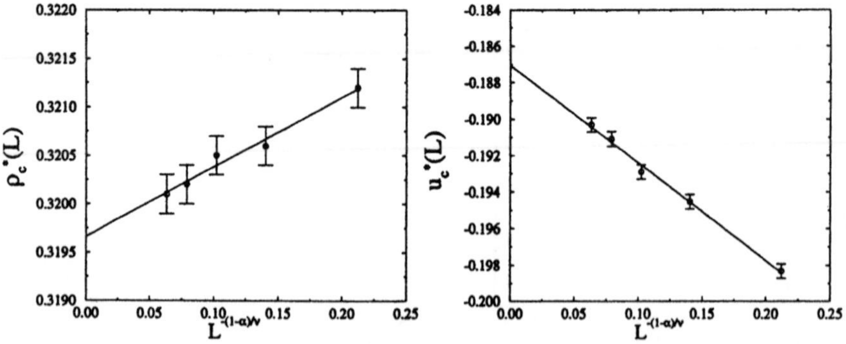

Figure 8: (a) The measured average density $\langle\rho\rangle_c(L)$ at the designated critical point, expressed as a function of $L^{-(1-\alpha)/\nu)}$. The least-squares fit yields an infinite volume estimate $\rho_c = 0.3197(4)$. (b) The measured average energy density $\langle u\rangle_c(L)$ at the critical point, expressed as a function of $L^{-(1-\alpha)/\nu}$. Extrapolation of the least-squares fit to infinite volume yields the estimate $u_c = -0.187(2)$. In both cases we took $1/\nu = 1.5887$[4].

In figure 8 we plot the values of $\langle\rho\rangle_c$ and $\langle u\rangle_c$, corresponding to the distributions of figure 7, as a function of $L^{-(1-\alpha)/\nu}$. Although no allowances have been made for corrections to scaling (the effects of which are certainly much smaller than those of field mixing), the data exhibit, within the uncertainties, a rather clear linear dependence. Least-squares fits to the data yield the infinite volume estimates $\rho_c^* = 0.3197(4)$, $u_c^* = -0.187(2)$.

4 Conclusions

In summary, we have demonstrated the manner in which field mixing influences the non-universal critical properties of fluids. In particular it has been shown that the FSS behaviour of the order parameter and energy distributions in critical fluids is very different from the Ising model. In fluids, the FSS scaling counterparts of the Ising order parameter and energy are the respective scaling operators $\mathcal{M}$ and $\mathcal{E}$, both of which are linear combinations of the fluid order parameter and energy. By effecting the data collapse of $p_L(\mathcal{M})$ onto the critical Ising magnetisation distribution, highly accurate determinations of fluid critical point parameters are possible. This use of the scaling operator distributions represents the natural extension to fluids of the order parameter distribution methods previously deployed so successfully in studies of symmetric spin models. Thus, in principle at least, there would appear to be no barriers to attaining similar degrees of accuracy in the study of critical fluids as is routinely achieved for lattice spin systems.

Finally we remark that the techniques deployed here are not restricted to simple fluids, but can be combined with configurational bias Monte Carlo methods [29] to facilitate accurate investigations of phase coexistence and critical phenomena in polymer systems. Such extensions are the subject of ongoing work.

Acknowledgements

Part of the work reported here evolved from fruitful collaboration with A.D. Bruce and M. Müller. The author acknowledges the financial support of a European Network fellowship (ERB CHRX CT-930 351).

References

[1] For a review, see V. Privman (ed.) *Finite size scaling and numerical simulation of statistical systems* (World Scientific, Singapore) (1990).

[2] For a review see K. Binder in *Computational Methods in Field Theory* H. Gausterer, C.B. Lang (eds.) (Springer-Verlag Berlin-Heidelberg) 59-125 (1992).

[3] K. Binder, Z. Phys. **B43**, 119 (1981).

[4] A.M. Ferrenberg and D.P. Landau, Phys. Rev. **B44**, 5081 (1991).

[5] D. Nicolaides and A.D. Bruce, J. Phys. A **21**, 233 (1988).

[6] A.P. Gottlob and M. Hasenbusch, Physica **A201**, 593 (1993).

[7] P. Peczak, A.M. Ferrenberg and D.P. Landau, Phys. Rev. **B43**, 6087 (1991).

[8] K. Chen, A.M. Ferrenberg and D.P. Landau, Phys. Rev. **B48**, 3249 (1993).

[9] M. Rovere, D.W. Heermann and K. Binder, J. Phys.: Condens. Matter **2**, 7009 (1990).

[10] M. Rovere, P. Nielaba and K. Binder, Z. Phys. **B90**, 215 (1993).

[11] A.D. Bruce and N.B. Wilding, Phys. Rev. Lett. **68**, 193 (1992).

[12] N.B. Wilding and A.D. Bruce, J. Phys.: Condens. Matter **4**, 3087 (1992).

[13] N.B. Wilding, Z. Phys. **B93**, 119 (1993).

[14] M. Müller and N.B. Wilding, Phys. Rev. E (in press).

[15] J.V. Sengers and J.M.H. Levelt Sengers, Ann. Rev. Phys. Chem. **37**, 189 (1986).

[16] F.J. Wegner, Phys. Rev. **B5**, 4529 (1972).

[17] A.D. Bruce, J. Phys. C **14**, 3667 (1981).

[18] J.J. Rehr and N.D. Mermin, Phys. Rev. A **8**, 472 (1973).

[19] For an analytic calculation of this function see R. Hilfer and N.B. Wilding, Mainz preprint.

[20] The universal function representing the *difference* between the finite-size form of $p_L(\rho)$ and its limiting form $\tilde{p}^{\star}_{\mathcal{M}}(x)$ has been previously studied in references [12, 13], where it was termed the 'field mixing correction'.

[21] N.B. Wilding and M. Müller, J. Chem. Phys. (in press).

[22] For a fuller report of this work see N.B. Wilding, Mainz preprint.

[23] D.J. Adams, Mol. Phys. **29**, 307 (1975).

[24] M.P. Allen and D.J. Tildesley *Computer simulation of liquids* (Oxford University Press,Oxford) (1987).

[25] B. Smit, J. Chem. Phys. **96** 8639 (1992); B. Smit and D. Frenkel, J. Chem. Phys . **96** 5663 (1991).

[26] A.Z. Panagiotopoulos, preprint.

[27] A.M. Ferrenberg and R.H. Swendsen, Phys. Rev. Lett. **61** 2635 (1988); A.M. Ferrenberg and R.H. Swendsen Phys. Rev. Lett. **63**, 1195 (1989). R.H. Swendsen, Physica A**194**, 53 (1993).

[28] B.G. Nickel and J.J. Rehr, J. Stat. Phys. **61**, 1 (1990).

[29] J.I. Siepmann, Mol. Phys. **70**, 1145 (1990).

Growth and Etching of GaAs(001)

P. Šmilauer[1]**and D.D. Vvedensky*[2]

[1]Interdisciplinary Research Center for Semiconductor Materials, The Blackett Laboratory, Imperial College, London SW7 2BZ, UK
[2]Department of Physics, The Blackett Laboratory, Imperial College, London SW7 2BZ, United Kingdom
(electronic address: d.vvedensky@ic.ac.uk)

Abstract. Recent progress in the atomic–scale modeling of epitaxial growth and chemical–beam etching of GaAs(001) using kinetic Monte Carlo simulations of a simple solid–on–solid model is reviewed. The additional activation barrier to interlayer transport emerges as a key factor for the evolution of the surface morphology during both growth and etching.

1. Molecular beam epitaxy and chemical–beam etching of GaAs(001)

Homoepitaxial growth of GaAs by molecular–beam epitaxy (MBE) is of indisputable practical significance [1] and is frequently used in fundamental studies of epitaxial growth. In principle, the GaAs MBE methodology is simple: molecular beams of Ga and As_2 (or As_4) are directed to a heated crystalline GaAs substrate placed in an ultrahigh–vacuum chamber. However, the details of the growth process and surface chemistry are complex and largely unknown.

The evolving surface morphology, usually monitored with reflection high–energy electron diffraction (RHEED), is a result of a competition between the randomness induced by the fluctuations in the flux of incoming beams and smoothing and self–organization due to surface diffusion. Under favorable growth conditions, the smooth surface morphology required in applications can be achieved. The principal goal of theoretical research in MBE is to identify such growth conditions and, more generally, to obtain detailed information on the growth process, which would allow the reproducible fabrication of atomic–scale semiconductor heterostructures.

Very recently, the removal of atoms during processes such as sputtering and etching has also attracted considerable attention as a way of *in situ* manipulation of surface morphology. In particular, etching of GaAs using molecular beams of trihalides (PCl_3, $AsCl_3$) was demonstrated [2] to provide monolayer etching precision while being fully compatible with the MBE experimental set-up. Once again, the details of the chemical reactions during the etching process are unknown. From a theoretical standpoint, the removal of material is the inverse process to growth, with surface monovacancies and vacancy islands playing the roles of adatoms and adatom islands, respectively.

In the following, we summarize recent results obtained with a very simple model of growth and etching of GaAs(001) substrates and give a description of the refinements required to study these phenomena. It is found that in all the cases studied, a prominent role is played by step–edge barriers to interlayer atomic transport.

---

*Present address: HLRZ, Forschungszentrum Jülich, 52425 Jülich, Germany (electronic address: pavel@hlrserv.hlrz.kfa-juelich.de); on leave from Institute of Physics, Cukrovarnická 10, 162 00 Praha 6, Czech Republic.

Springer Proceedings in Physics, Vol. 80
Computer Simulation Studies in Condensed-Matter Physics VIII
Editors: D.P. Landau · K.K. Mon · H.-B. Schüttler © Springer-Verlag Berlin Heidelberg 1995

2. Kinetic Monte Carlo model

Kinetic Monte Carlo (KMC) simulations of a solid-on-solid (SOS) model are at present the most versatile and powerful tool for investigating both epitaxial growth and the removal of material.

The basic variant of the model we use was proposed as a direct descendant of the SOS model developed in the 1970s by Gilmer and co-workers [3] in which the crystal substrate is treated as a simple cubic lattice with neither bulk vacancies nor overhangs allowed. Clarke and Vvedensky [4] adopted a strategy very different from previous modeling work on growth of III–V compounds [5]. Instead of trying to replicate complicated details of GaAs(001) surface chemistry, they developed a model for simulating growth under typical growth conditions, i.e., with sufficiently large As overpressure [1]. The model is thus based on the assumption that As incorporation kinetics are not a rate–limiting step of the growth process and need not be included explicitly in its description. The model can be thought of as focusing on the Ga incorporation while the effects of As incorporation and other fast processes, as well as specific features of the GaAs(001) surface such as the existence of surface reconstructions and the resulting anisotropy in surface diffusion and sticking coefficients, affect only the *effective* model parameters. This strategy allowed the authors of Ref. [4] to develop a model with only two free parameters while retaining the essential features of the growth kinetics. It is important to bear in mind that such a simplification restricts the applicability of the model (e.g., the effects resulting from variations of As pressure cannot be straightforwardly investigated) but this is outweighed by benefits of having only a few model parameters and higher computational efficiency.

Two kinetic processes were included in the original model to describe growth (see Fig. 1): random deposition of atoms onto the substrate and migration of surface adatoms. Desorption is neglected in accordance with experimental observations [6]. Surface migration is modeled as a nearest–neighbor hopping process, with the hopping rate of the Arrhenius form, $k(E,T) = k_0 \exp(-E/k_BT)$, where k_0 is the attempt frequency (usually taken to be equal to $2k_BT/h$), T is the substrate temperature, k_B is Boltzmann's constant, h is Planck's constant and E is the energy barrier to hopping. The barrier E depends on the adatom environment and comprises a substrate term, E_S, and a contribution, E_N, from each lateral nearest neighbor in the original position of the hopping atom (cf. Fig. 1). Note that the growth of GaAs was found to be best modeled using $E_S \gg E_N$. The use of the Arrhenius barriers with the hopping

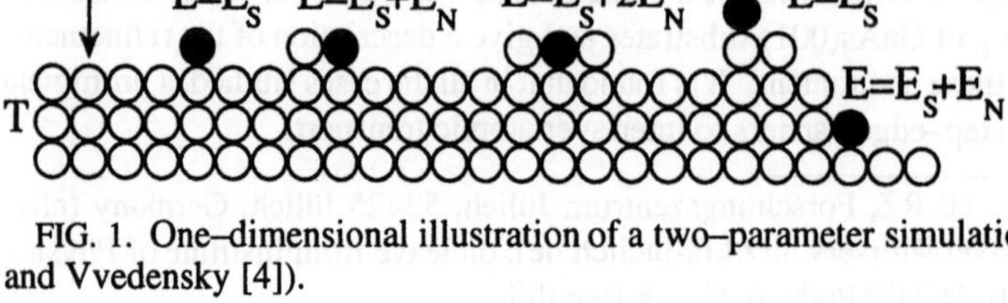

FIG. 1. One–dimensional illustration of a two–parameter simulation model (Clarke and Vvedensky [4]).

barrier depending on the atom coordination n in its initial position, $E = E_S + nE_N$, is motivated by the observation [7] that other popular choices in which the barrier depends on both the initial and final site coordination (such as Metropolis or Kawasaki dynamics) are faster in achieving thermal equilibrium but may lead to unphysical evolution of the system *towards* equilibrium and thus are not suitable for modeling far–from–equilibrium growth kinetics. There is also overwhelming experimental [8] and theoretical [9] evidence suggesting that in many cases the surface diffusion processes do obey the Arrhenius (barrier–crossing) dynamics.

Another important feature of the approach adopted in Ref. [4] concerns the modeling of the evolution of the RHEED specular–beam intensity during growth. While RHEED is in principle a straightforward experimental technique, its theoretical treatment is probably the most difficult of all surface–sensitive diffraction techniques currently used. It is precisely the very small angle of incidence of the electron beam (typically 0.5–3°) making the method easily compatible with MBE experimental set-up that leads to increased importance of multiple scattering and surface reconstruction effects. Small changes of an azimuth or a polar angle may lead to changes in the amplitude, phase, and even in period of the RHEED oscillations [10] which, to the best of our knowledge, has not been so far successfully explained by any theory. In fact, "in–phase", or "Bragg" diffraction conditions are often used in experiments, for which the kinematic theory yields a constant intensity regardless of the surface configuration, in contrast to the experimentally observed strong oscillations. Therefore, the density of surface steps was used in Ref. [4] to model the *variations* of the RHEED specular–beam intensity during growth.

Below, we describe the extensions of the original model which we have adopted to study growth and etching of GaAs(001). Included are, e.g., the implementation of the additional barriers to hopping at step edges, the process of incorporation of an incoming atom into the growing crystal and modeling of faster vacancy diffusion.

3. Growth on a vicinal surface

Growth on vicinal surfaces, with a misorientation angle of few degrees with respect to a high–symmetry orientation of the substrate, is of profound importance as a method of preparing high–quality epitaxial layers and heterostructures and is also frequently used to obtain insight into the fundamental kinetics of MBE growth [6,11]. A vicinal surface consists of a train of terraces with some average length – a new characteristic length for the surface diffusion kinetics. On a *singular* surface at a high substrate temperature, growth typically proceeds in a layer-by-layer mode in which layers are periodically formed by the nucleation, growth, and coalescence of adatom islands. This is accompanied by oscillations of the intensity of a specular RHEED beam. In contrast, on a *vicinal* surface at a high enough substrate temperature, most of the atoms deposited on the surface reach terrace edges and the growth proceeds by advancement of steps (the step–flow growth regime frequently used in applications to produce a very smooth surface morphology). If the temperature of the substrate is decreased, however, the diffusion length of free adatoms decreases exponentially and at a transition temperature, for which the average terrace length becomes equal to the *effective* diffusion length, the growth mode changes to layer–by–layer. This transition between different growth modes can be used to estimate the surface diffusion constant [6] though this is by no means a trivial task [12].

In the experiments reported by Shitara *et al.* [11], the transition to the step–flow growth mode was used to parameterize the KMC model of Ref. [4]. Rather at variance with common practice, the growth and diffraction conditions in these experiments were chosen to conform as closely as possible to the assumptions of this simulation model. The As/Ga (atomic) ratio was held at approximately 2.5 to maintain the 2×4 surface reconstruction over a fairly wide temperature range near the temperature at which growth becomes dominated by step flow. This ensures that the As incorporation kinetics are not the rate–limiting step in the growth process and also that the effect of the surface reconstruction on the adatom mobility and RHEED intensity is unchanged during growth at different temperatures. The effect of surface anisotropies, in particular the anisotropy of the sticking coefficients at [110] and [$\bar{1}$10] steps, was diminished by the use of the surfaces misoriented toward the [010] direction on which both types of steps are present in approximately equal proportion. In–phase diffraction conditions were chosen for which the maxima in the specular–beam intensities on the corresponding singular surface occur at monolayer increments of material deposited. As a result of fitting the model parameters (E_S and E_N), *quantitative* agreement was achieved such that relative changes of the magnitudes of the RHEED specular–beam intensity and the step density in simulations, for different growth rates and substrate misorientations, were the same during growth.

On the other hand, there was also clear disagreement during post–growth recovery (smoothing of the surface which follows termination of the Ga flux while the As_2 flux and the substrate temperature are maintained), which was most evident at lower substrate temperatures. The simulations generally showed a much more rapid recovery than the measured RHEED profiles (cf. Ref. [11]) and were thus unable to reproduce the systematic temperature dependence of the time constant for the initial stage of the recovery (see Refs. [13] and [14] for detailed discussions). This is a serious discrepancy because the equilibration of a surface after a period of growth is a more discriminating test of kinetic models than growth alone. During growth under typical far–from–equilibrium conditions, the maximum time scale is set by the deposition flux and processes occurring over longer time scales are "frozen out". However, during recovery, these processes can come into play even though their effect during growth can be safely omitted [14].

To correctly model the equilibration of the surface, we assumed that there are activation barriers to hopping at step edges on GaAs(001) surfaces [15]. Note that there was no direct evidence at that time that such barriers might be present on a semiconductor surface although they had been for a long time known to be present on some metal surfaces [16] and their effect on growth on vicinal surfaces had been theoretically investigated [17,18]. These barriers are expected to inhibit the interlayer transport of adatoms, thus slowing down the recovery [19]. A convenient implementation of the additional (step–edge) barrier to hopping is to count the number of next–nearest neighbors in the planes beneath and above the hopping atom before (m_i) and after (m_f) a hop. The barrier has a nonzero value only if $m_i > m_f$ and equals $(m_i - m_f)E_B$ where E_B is a model parameter. The total energy barrier E is thus given

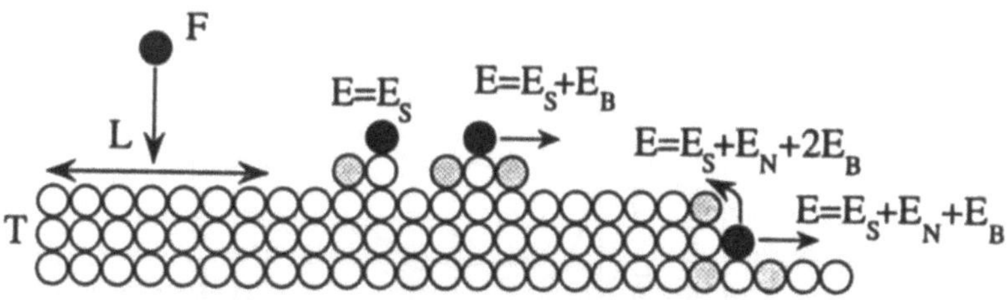

FIG. 2. One–dimensional illustration of the simulation model with the step–edge barrier and the post–deposition incorporation process.

by $E = E_S + nE_N + (m_i - m_f)\theta(m_i - m_f)E_B$ where $\theta(x) = 1$ if $x > 0$ and $\theta(x) = 0$ if $x \leq 0$ (see Fig. 2).

However, we found that even quite a small step–edge barrier has important consequences during growth as well. The surface quickly roughens and complete disagreement between simulations and experiment during growth *and* recovery is observed. We were thus led to the conclusion that if step–edge barriers are present then some compensating smoothing mechanism must also be present. We have therefore included an additional process that an arriving atom undergoes before being incorporated on the surface. A search is carried out within a square of a fixed linear size L, centered upon the original site, for the site that maximizes the number of nearest neighbors. The effect of this process is a smoothing of edges of pre–existing steps and islands created on the terraces, but we can only speculate about its possible microscopic origins for the case of a GaAs(001) surface. The simulations were carried out on 232×232 lattices with steps running diagonally across the lattice (a miscut toward [010] direction) [11] with the parameters $E_S = 1.54$ eV, $E_N = 0.23$ eV, $E_B = 0.175$ eV and $L = 7$ and with the substrate temperatures and the growth rates as used in experiments.

Fig. 3 shows a good agreement between the RHEED specular–beam intensity and the surface step density over the entire growth and recovery period. This agreement is compelling not least because the effects of the incorporation process and the step–edge barrier act in *opposition* during growth, as discussed above, but in *concert* during recovery. The step–edge barrier slows down the equilibration process by inhibiting interlayer transport, while the incorporation process produces smoother step edges and a decrease in the number of free adatoms. Most of the surface atoms are high–coordinated, which reduces the influence of fast processes such as free adatoms migrating to high–coordinated sites and the elimination of sites with low coordination, both of which were over–emphasized in the original model. Therefore, according to our simulations, both step-edge barriers and the incorporation process are important factors for correctly reproducing the recovery curves.

The incorporation mechanism is responsible for yet another improvement of agreement between experimental results and simulations. The close inspection of Fig. 3 reveals a gradual delay of the first maximum of the RHEED oscillations as the temperature is increased. This phenomenon is caused by an increasing proportion of atoms being captured by terrace edges and thus not participating in nucleation of islands on terraces [20]. Therefore, more atoms have to be deposited to complete a layer and

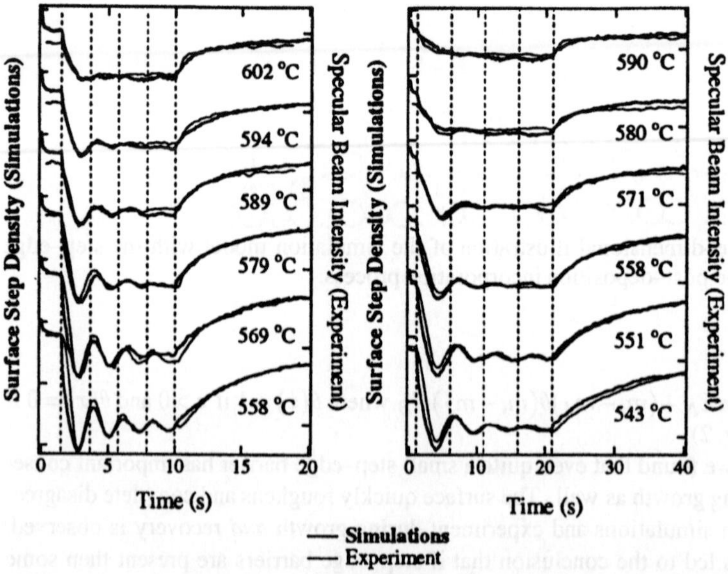

FIG. 3. Direct comparison between measured RHEED intensity and the simulated surface step density on a GaAs(001) vicinal surface misoriented by 2° toward [010] direction at the growth rate of 0.47 ML/s (left panel) and 0.20 ML/s (right panel) [15].

the first maximum of the RHEED intensity shifts to longer times. The first maximum delay in the original simulations [11] was much smaller than observed experimentally. The reason why the present model is more successful is the use of the incorporation mechanism: in the original model, there was an appreciable probability for (typically) low–coordinated atoms to detach from the terrace edge and return on the terrace to participate in island nucleation and growth. This process is strongly suppressed in the extended version of the model because the terrace edges are very compact and detachment of the high–coordinated edge atoms has a very low probability.

4. Growth on a singular surface

The step–edge barrier has a stabilizing effect on step–flow growth on vicinal surfaces, as discussed many years ago [17], because it causes most of the atoms to remain on the terrace on which they were originally deposited. Thus, the steps fed from longer terraces advance faster compared to those supplied by atoms landing on short terraces, which leads to equalization of terrace lengths. However, the situation is very different on a singular surface (or on a vicinal surface with the average terrace length much larger than the effective diffusion length of adatoms) where the growth becomes asymptotically unstable [21]. The hindered interlayer transport causes earlier nucleation on top of islands and makes it difficult to complete lower layers. As a result, large pyramid–like features (mounds) develop on the surface as recently observed

[22,23] for growth on a nominally flat GaAs(001) surface using scanning tunneling and atomic force microscopy.

We performed simulations on 300×300 lattices with the same model parameters as used in the simulations of recovery on GaAs(001), and at the substrate temperature 828 K and the growth rate 1/6 ML/s as used in the experiments of Refs. [22] and [23]. We observed the development of mounds on the growing singular surfaces after several tens of monolayers were deposited and the initial oscillations of the step density have died out. The mounds in our simulations grow and merge but occasionally also split, as shown in Fig. 4 where the mound marked by the arrow splits into two mounds which again merge after an additional several tens of monolayers are grown. Finally, the surface arrives at a state in which the number of mounds and their sizes change very slowly during deposition of several hundred monolayers, although the coarsening process still proceeds and the shape of the mounds becomes more regular (cf. Fig. 4). Throughout the simulation, the mounds keep an approximately constant average angle of inclination, in qualitative agreement with the experiment. Both in the experiments and in our simulations the size of the terraces on the hillsides of the mounds and the average inter-island distance at submonolayer coverages are roughly the same [24]. We also simulated equilibration of the mounds after the growth was stopped and observed a very slow approach to a flat surface, with the residua of mounds distinguishable even after 1000 s annealing, again in very good agreement with the experiment [23].

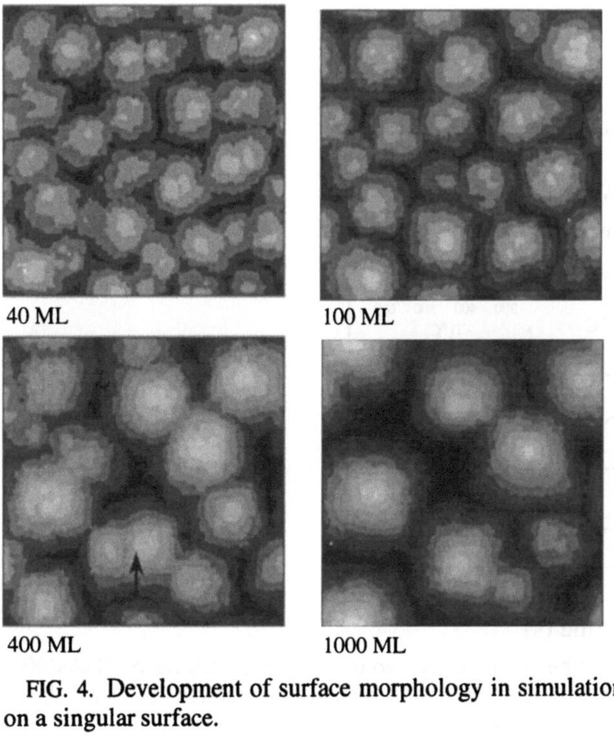

FIG. 4. Development of surface morphology in simulations of GaAs(001) growth on a singular surface.

5. Etching of a singular surface

The removal of material with a monolayer precision is of great practical importance. Recent experiments on etching of a singular GaAs(001) surface by $AsBr_3$ [25] have shown that exciting physical effects can result as well.

The experimental set-up for chemical–beam etching is very similar to that for MBE. A beam of $AsBr_3$ and a concomitant As_2 beam are directed onto a heated GaAs substrate and the evolution of the surface morphology is monitored using RHEED. In Ref. [25], a series of etching experiments at different substrate temperatures but with otherwise identical experimental conditions is reported. The evolutions of RHEED specular beam intensities from this study are shown in Fig. 5. A "re-entrant" behavior of the oscillations is observed, with sustained specular–beam intensity oscillations at high substrate temperatures, a decaying intensity with no oscillations at intermediate temperatures, but oscillations reappearing at still lower temperatures. Such re-entrant oscillatory behavior has to the best of our knowledge not been observed in any study of material removal, although re-entrant layer–by–layer *growth* was reported by Kunkel *et al.* [26] for homoepitaxy on Pt(111). Note that the temperature range over which the oscillations disappear is very narrow, extending only over $\approx 10°$C. The RHEED pattern in this range shows well–defined spots soon after etching commences, which indicates transmission through three–dimensional asperities on the surface. Above and below this temperature range, the RHEED pattern is streaked over the duration of the measurement.

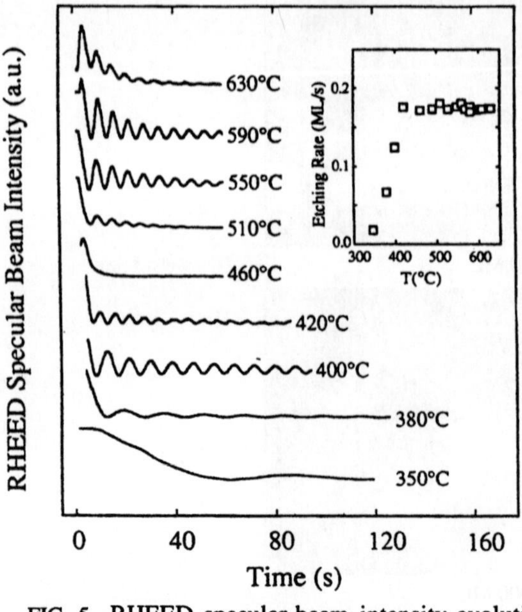

FIG. 5. RHEED specular-beam intensity evolution during etching of GaAs(001) with an $AsBr_3$ flux of 0.36 sccm and different substrate temperatures. The inset shows the etching rate vs. the substrate temperature.

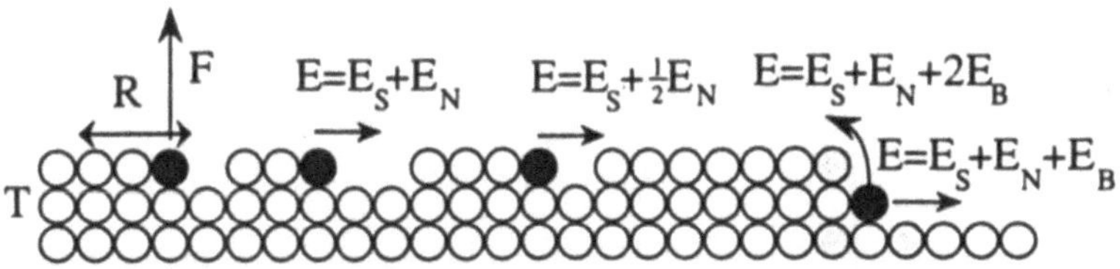

FIG. 6. One–dimensional illustration of the model used for simulation of chemical–beam etching, with the step–edge barrier, increased rates of vacancy and edge diffusion, and the preferential removal of low–coordinated atoms.

The inset to Fig. 5 shows the substrate temperature dependence of the etching rate as deduced from the period of the RHEED intensity oscillations. This rate is nearly constant above 420°C but diminishes rapidly below this temperature. We can conclude that above 420°C the removal is *supply–rate limited* and below 420°C it is *reaction–rate limited*. The important conclusion is that the re-entrant behavior is not due to any pathology in the $AsBr_3$ decomposition rate over the temperature range of interest, but rather to factors involving surface morphology and the kinetics of the atomistic processes.

To model the observations summarized in Fig. 5, we modified our KMC model in the following way (see Fig. 6). First, the random deposition process of our growth simulations is replaced by removal of randomly selected surface atoms. We include a site selectivity whereby atoms with low coordination are preferentially removed from the surface: a site is randomly selected, and then a search is carried out within a square of a fixed linear size R, centered at the originally selected site, for the surface atom with the fewest lateral nearest neighbors. This atom is then removed from the surface.

The step–edge barrier is included in the model in the same way as described above. However, the original nearest-neighbor contribution was modified to achieve realistic vacancy mobilities consistent with experimental observations [27–29]. Instead of simply counting the nearest neighbors in the original position, the nearest–neighbor contribution is now given by $n_i E_N/2+(n_i-n_f)\theta(n_i-n_f)E_N/2$ where n_i ($\equiv n$) and n_f are the number of lateral nearest neighbors before and after a hop, respectively. This definition increases the rate of diffusion along island edges and leads to a high vacancy mobility but preserves the difference in the hopping rate of edge atoms with different coordination as well as the activation barriers to detachment of atoms from step edges (the nearest–neighbor contribution is equal to $n_i E_N$ for $n_f=0$), cf. Fig. 6. The simulations were carried out on 200×200 lattices, the parameter R was set equal to 3.

The results of the simulation are shown in Fig. 7. The re-entrant nature of the RHEED oscillations is seen to be reproduced by the behavior of the step density in the simulations. Notice that the ranges of temperatures at which the oscillations are observed correspond well to those in Fig. 5. The temperature range over which the oscillations disappear (not shown) is very narrow, in qualitative agreement with the experiment.

Based on the experimental measurements and our simulations, the explanation of the re-entrant etching oscillations is as follows. Etching creates surface vacancies which

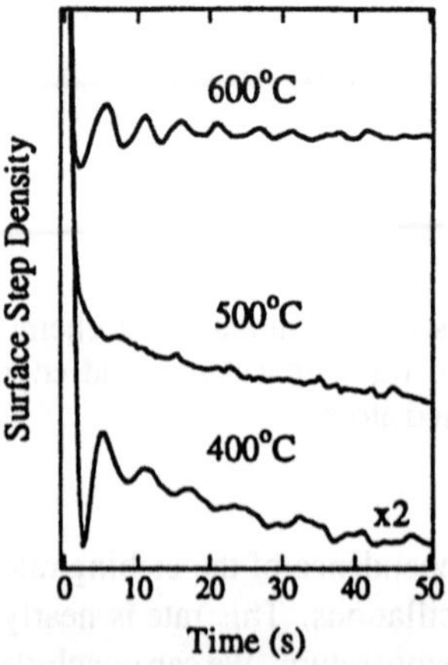

FIG. 7. The surface step density during growth at three different substrate temperatures. The step density increases *downwards*. The upper two curves in both plots are given offsets to make comparisons easier.

diffuse on the surface and form islands. As these islands grow, new vacancies are created in lower layers within the flat regions near their centers. At high temperatures ($\gg 460°$C), there are many free adatoms on the surface. Due to the high thermal mobility of adatoms and vacancies, the vacancies created in lower layers are filled in [30]. This results in the periodic removal of individual layers through the formation, growth, and coalescence of vacancy islands. The interlayer atomic transport in this temperature range is appreciable because the high density of free adatoms dominates the inhibiting effect of the step–edge barrier.

At intermediate temperatures ($\approx 460°$C) the number of free adatoms on the surface is reduced drastically because the rate of their detachment from step edges decreases exponentially with decreasing temperature. The filling in of vacancies created in lower layers is now strongly hindered by the step–edge barrier and, therefore, vacancy islands in lower layers are formed. These islands are nested inside vacancy islands on higher layers, creating pits (inverse pyramids) in the surface (cf. the scanning–tunneling microscopy images of a sputtered Pt(111) surface in Ref. [28]). As a result of this *multilayer*, or *three–dimensional*, mode of removal, a rough surface morphology quickly develops which is qualitatively different from both the low– and high–temperature morphologies that are essentially smooth. The multilayer removal is accompanied by the disappearance of RHEED oscillations and the appearance of a spotty diffraction pattern.

Finally, at low temperatures ($\ll 460°$C), the mobility of vacancies is low, which leads to the formation of many small islands and a high density of atoms with low coordination in the uppermost layer. These atoms are preferentially removed by the etching process which leads to the re-emergence of an approximately layer–by–layer removal mode and the reappearance of RHEED oscillations. The transition from the layer-by-layer removal regime at high temperatures to the multilayer removal regime has been observed in the ion-beam sputtering of Pt(111) [27,28], but obtaining the reappearance of the oscillations at lower temperatures requires a site-selective step such as the one we have proposed.

There is an important consequence of our interpretation of the reappearance of RHEED oscillations at low temperatures. If the etching rate is increased in the intermediate temperature regime by supplying more etchant molecules to the substrate, the oscillations should reappear, as if the substrate temperature had been decreased with the original flux maintained. In both cases the vacancy islands become smaller, the density of low-coordinated atoms at the uppermost layer is higher and the site-selective step becomes more important. The measurements and simulations [25] demonstrate that this is indeed the case: as the etching rate is increased, oscillations reappear at 460°C and their amplitude grows with the etching rate. This result demonstrates once more that the re-entrance of oscillations involves the surface morphology and not the change in the etching reaction with the temperature.

6. Summary and outlook

What makes the KMC model described above successful in describing growth and etching of GaAs(001) despite all its simplifications? We believe there are two basic reasons. First, the physical features of the system under study that are essential for reproducing the experimental observations are correctly captured by the model. In particular, given the quantitative agreement between RHEED specular–beam intensity and the surface step density in simulations, Ga diffusion and incorporation emerge as the key processes that determine growth kinetics under typical growth conditions. Second, many of the questions we try to answer using the model are about phenomena that are universal, insensitive to specific system details – similar phenomena are observed for other materials and different processes. For example, mounds develop and coarsen during growth on high–symmetry surfaces of metals as well [31,32], and the transition from layer–by–layer to multilayer removal was observed on a Pt(111) surface sputtered with low–energy ions [27,28].

It is obvious, however, that generalizations and extensions of the model are required to address other issues. Indeed, we did not discuss, e.g., the inclusion of anisotropy [11] or next–nearest–neighbor interactions [33] into the model. For detailed studies of more complex phenomena, a new approach will be required. A combination of quantum–mechanical total energy calculations (either approximate or *ab–initio*) and KMC modeling seems to offer a promising avenue. The energy barriers and details of atomistic processes obtained from the total energy calculations can be used in KMC simulations to model large–scale and long–time evolution of the system [9,34].

7. Acknowledgements

We would like to thank T. Shitara for providing us with his experimental data used in Fig. 3. The support of Imperial College and the Research Development Corporation of Japan is gratefully acknowledged.

References

[1] *Molecular beam epitaxy*, edited by A. Cho (American Institute of Physics, New York, 1994).

[2] W.T. Tsang, R. Kapre, and P.F. Sciortino, Appl. Phys. Lett. **62**, 2084 (1993); W.T. Tsang, T.H. Chiu, and R.M. Kapre, Appl. Phys. Lett. **63**, 3500 (1993).

[3] G.H. Gilmer and P. Bennema, J. Appl. Phys. **43**, 1347 (1972); J.D. Weeks and G.H. Gilmer, Adv. Chem. Phys. **40**, 157 (1979).

[4] S. Clarke and D.D. Vvedensky, Phys. Rev. Lett. **58**, 2235 (1987).

[5] See A. Madhukar and S.V. Ghaisas, CRC Critical Reviews in Solid State and Materials Science **14**, 1 (1988) for a review.

[6] J.H. Neave, P.J. Dobson, B.A. Joyce, and J. Zhang, Appl. Phys. Lett. **47**, 100 (1985).

[7] H.C. Kang and W.H. Weinberg, J. Chem. Phys. **90**, 2824 (1989).

[8] A. Zangwill, *Physics at Surfaces* (Cambridge University Press, Cambridge, England 1988).

[9] Z. Zhang, Y.-T. Lu, and H. Metiu, Surf. Sci. **255**, L543 (1991); C.-L. Liu and J.B. Adams, Surf. Sci. **265**, 262 (1992).

[10] J. Zhang, J.H. Neave, P.J. Dobson, and B.A. Joyce, Appl. Phys. A **42**, 317 (1987).

[11] T. Shitara, D.D. Vvedensky, M.R. Wilby, J. Zhang, J.H. Neave, and B.A. Joyce, Phys. Rev. B **46**, 6815 (1992); *ibid.* **46**, 6825 (1992).

[12] A.K. Myers–Beaghton and D.D. Vvedensky, Phys. Rev. B **42**, 5544 (1990); S. Stoyanov, Appl. Phys. A **50**, 349 (1990).

[13] J.H. Neave, B.A. Joyce, P.J. Dobson, and N. Norton, Appl. Phys. A **31**, 1 (1983); F.J. Grunthaner, A. Madhukar, T.C. Lee, and R. Fernandez, J. Vac. Sci. Technol. B **3**, 1317 (1985); A. Yoshinaga, M. Fahy, S. Dosanjh, J. Zhang, J.H. Neave, and B.A. Joyce, Surf. Sci. **264**, L157 (1992).

[14] D.D. Vvedensky and S. Clarke, Surf. Sci. **225**, 373 (1990).

[15] P. Šmilauer and D.D. Vvedensky, Phys. Rev. B **48**, 17603 (1993).

[16] G. Ehrlich and F.G. Hudda, J. Chem. Phys. **44**, 1039 (1966).

[17] R.L. Schwoebel and E.J. Shipsey, J. Appl. Phys. **37**, 3682 (1966); R.L. Schwoebel, J. Appl. Phys. **40**, 614 (1969).

[18] S.G. Bales and A. Zangwill, Phys. Rev. B **41**, 5500 (1990).

[19] L.–M. Peng and M.J. Whelan, Proc. R. Soc. London **435**, 257 (1991).

[20] H.T.W. Zandvliet, H.B. Elswijk, D. Dijkkamp, E.J. van Loenen, and J. Dieleman, J. Appl. Phys. **70**, 2614 (1991); T. Shitara, J. Zhang, J.H. Neave, and B.A. Joyce, J. Appl. Phys. **71**, 4299 (1992).

[21] J. Villain, J. Phys. I (France) **1**, 19 (1991).

[22] M.D. Johnson, C. Orme, A.W. Hunt, D. Graff, J.L. Sudijono, L.M. Sander, and B.G. Orr, Phys. Rev. Lett. **72**, 116 (1994).

[23] C. Orme, M.D. Johnson, J.L. Sudijono, K.T. Leung, and B.G. Orr, Appl. Phys. Lett. **64**, 860 (1994).

[24] P. Šmilauer and D.D. Vvedensky (unpublished).

[25] T. Kaneko, P. Šmilauer, B.A. Joyce, T. Kawamura, and D.D. Vvedensky (unpublished).

[26] R. Kunkel, B. Poelsema, L.K. Verheij, and G. Comsa, Phys. Rev. Lett. **65**, 733 (1990).
[27] B. Poelsema, L.K. Verheij, and G. Comsa, Phys. Rev. Lett. **53**, 2500 (1984).
[28] T. Michely, T. Land, U. Littmark, and G. Comsa, Surf. Sci. **272**, 204 (1992).
[29] P. Bedrossian, J.E. Houston, J.Y. Tsao, E. Chason, and S.T. Picraux, Phys. Rev. Lett. **67**, 124 (1991); P. Bedrossian and T. Klitsner, Phys. Rev. Lett. **68**, 646 (1992); P. Bedrossian, Surf. Sci. **301**, 223 (1994).
[30] Possible mechanisms for the annihilation of vacancies formed in lower layers and the role of adatom and vacancy migration are discussed in Refs. [28] and [35].
[31] M. Bott, T. Michely, and G. Comsa, Surf. Sci. **272**, 161 (1992).
[32] J.A. Stroscio, D.T. Pierce, M. Stiles, A. Zangwill, and L.M. Sander (unpublished).
[33] N. Haider, M.R. Wilby, and D.D. Vvedensky, Appl. Phys. Lett. **62**, 3108 (1993).
[34] A.F. Voter, Phys. Rev. B **34**, 6819 (1986).
[35] P. Šmilauer, M.R. Wilby, and D.D. Vvedensky, Surf. Sci. Lett. **291**, L733 (1993).

Monte-Carlo Simulations of Phase Transitions in Frustrated Antiferromagnets

M.L. Plumer[1] *and A. Mailhot*[2]

[1]Département de Physique, Université de Sherbrooke, Sherbrooke, Québec, Canada J1K 2R1
[2]INRS-EAU, 2800 rue Einstein, C.P. 7500, Sainte-Foy, Québec, Canada G1V 4C7

Abstract. Recent finite-size scaling results of histogram Monte-Carlo simulations on the stacked triangular antiferromagnet are reviewed. Estimates for critical exponents associated with phase transitions to novel symmetry states of the Ising, *XY*, and Heisenberg models are compared with theory and experiment. New results are presented for a magnetic-field induced 3-state Potts transition in the *XY* model which is very weakly first order in character.

1. Introduction

For some twenty years, critical exponents calculated by renormalisation-group (RG) methods (e.g., in $4-\epsilon$) were considered the standard for comparison with experimental results, as well as a variety of other theoretical techniques such as Monte Carlo (MC) simulations. Recent interest in frustrated systems which exhibit more complicated types of magnetic (and other) order has made clear that methods based on perturbation expansions can be unreliable in many cases [1]. It is for this reason, along with recent progress in computational techniques, that MC simulations are gaining importance in the understanding of the critical behavior of such systems.

Analysis of MC data by the histogram method has proven particularly useful in the estimation of critical temperatures as well as critical exponents and is well tested in a variety of unfrustrated systems [2]. The utility of this technique is that it allows for the determination of thermodynamic functions over a continuous range of temperature with a single simulation and is extremely useful for finite-size scaling (FSS) estimates of critical exponents as well as the detection of very weak first-order transitions. Although a number of new algorithms have been successful in reducing the effects of critical slowing-down, these appear to be largely ineffective for the frustrated systems of interest here [3].

The present work reviews our recent FSS results of histogram MC data for the stacked triangular (simple hexagonal) antiferromagnetic for the Ising, XY and Heisenberg models. (Earlier studies, mostly based on mean-field theory, are summarized in Refs. [4] and [5].) These systems are expected to exhibit unusual critical behavior of a controversial nature, which is also evident in a large number of experimental results [6]. Consideration of the spin density makes transparent the relevant symmetries involved and is useful in describing

Springer Proceedings in Physics, Vol. 80
Computer Simulation Studies in Condensed-Matter Physics VIII
Editors: D.P. Landau · K.K. Mon · H.-B. Schüttler

the structure of the long-range order. New results are also presented for a magnetic-field induced 3-state Potts transition in the XY model which is very weakly first order in character.

2. The Spin Density, Symmetries and Predictions

Spins on a triangular lattice with near-neighbor (NN) antiferromagnetic (AF) interactions cannot satisfy the the idealised up-down pairing of all bonds. In the case of vector spins, this *frustration* gives rise to the so-called 120° spin structure, as depicted in Fig. 1. Such types of non-colinear magnetic ordering are conveniently characterised by the spin density written as

$$\mathbf{s}(\mathbf{r}) = \mathbf{S}e^{i\mathbf{Q}\cdot\mathbf{r}} + \mathbf{S}^{*}e^{-i\mathbf{Q}\cdot\mathbf{r}}$$

where $\mathbf{S} = \mathbf{S}_a + i\mathbf{S}_b$ is the complex polarization vector and $\mathbf{Q}$ is the wavevector. In the case of ferromagnetic (F) interactions along the c axis, a modulation of period-3 occurs only in the basal plane and $\mathbf{Q} = (4\pi/3a)\hat{\mathbf{a}}$ (i.e., $3\mathbf{Q} = \mathbf{G}$, where $\mathbf{G}$ is a reciprocal lattice vector). However, for systems with AF NN interactions also along the c axis, an additional (unfrustrated) period-2 modulation occurs so that $\mathbf{Q} = (4\pi/3a)\hat{\mathbf{a}} + (\pi/c)\hat{\mathbf{c}}$ and the structure has an overall periodicity of 6 (i.e., $6\mathbf{Q} = \mathbf{G}$). The spin density can also be written as

$$\mathbf{s}(\mathbf{r}) = 2\mathbf{S}_a\cos(\mathbf{Q}\cdot\mathbf{r}) - 2\mathbf{S}_b\sin(\mathbf{Q}\cdot\mathbf{r})$$

which makes clear the role of frustration. Unfrustrated systems, such as bipartite lattices with NN AF interactions, have modulations characterised by $\mathbf{Q} = \frac{1}{2}\mathbf{G}$ so that the *sine* term above is always zero. Magnetic order thus involves a single polarization vector $\mathbf{S}_a$. In contrast, frustrated systems require both vectors $\mathbf{S}_a$ and $\mathbf{S}_b$ to describe the long-range order. This gives rise to a Landau-Ginsburg-Wilson Hamiltonian with *two* fourth-order terms, $u(\mathbf{S}\cdot\mathbf{S}^{*})^2 + v\mid \mathbf{S}\cdot\mathbf{S}\mid^2$, whereas only the first term is present in unfrustrated systems.

Several symmetry arguments based on these observations regarding the long-range ordering of the 3D frustrated Ising, XY and Heisenberg models have appeared over the past ten years. Berker *et al.* [7] noted that such an analysis leads to the expectation of XY universality associated with the phase

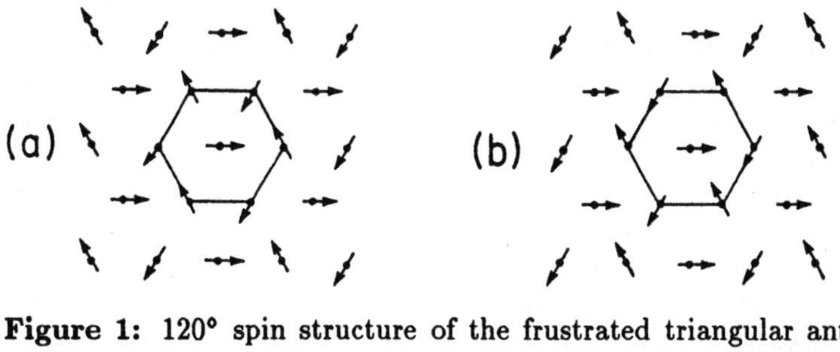

Figure 1: 120° spin structure of the frustrated triangular antiferromagnet. Shown are (a) left-handed and (b) right-handed degenerate chiral states.

transition for the *Ising* system since both (n=2) components of the spin density are involved in the ordering process. Recent MC simulation results, however, have raised the possibility of a new universality class [8]. In the case of vector spin models, the spin density above with *helical* polarization, $\mathbf{S}_a \perp \mathbf{S}_b$ and $S_a = S_b$ (i.e., $\mathbf{S} \cdot \mathbf{S} = 0$), describes the 120° spin structure. In contrast with the usual helimagnetic structure found in some rare-earth elements [5], [6] where $\mathbf{S} \perp \mathbf{Q}$, here the spin polarization vector lies in the same plane as the period-3 wavevector. Critical behavior is, however, expected to be independent of such considerations. For XY spins, in addition to the continuous rotational invariance $SO(2)$ also found in unfrustrated systems, there is a discrete two-fold (Z_2) *chiral* degeneracy so that the order parameter space has an overall symmetry $V = Z_2 \times SO(2)$ [29] (see Fig. 1). In the case of Heisenberg spins, the two orthogonal vectors map-out a solid sphere so that the symmetry of the order parameter is isomorphic to the three-dimensional rotation group $SO(3)$ [10].

Based on such arguments, as well as $4-\epsilon$ RG expansions and, in particular, results from standard MC-simulation analysis, Kawamura has argued in favor of new *chiral* universality classes for the frustrated XY and Heisenberg systems [11]. In contrast, the results of an analysis of the non-linear sigma model (appropriate for the Heisenberg case) in $2+\epsilon$ dimensions has led Azaria, Delamotte and Jolicoeur to the conclusion that these systems can exhibit nonuniversal critical behavior dependent upon unspecified microscopic parameters, where $O(4)$ universality, tricritical behavior, or a first-order transition will occur [1]. Zumbach [12] has also emphasized the possibility of an "almost second-order" phase transition with effective critical exponents (also considered by Kawamura). The most recent RG calculation suggests that both XY and Heisenberg systems have first-order transitions [13]. The long history of earlier RG analyses is reviewed in Refs. [1], [5], and [13]. Our recent FSS results discussed in the next section suggest XY universality, tricritical behavior, and the $n = 3$ chiral universality class for the Ising, XY, and Heisenberg models, respectively.

3. Finite-Size Scaling Estimates of Critical Exponents

For the three systems examined, the standard Metropolis algorithm was used to generate histograms on $L \times L \times L$ lattices. FSS analysis was made to estimate critical exponents of a number of thermodynamic quantities at the transition temperature, including the specific heat $C \sim L^{\alpha/\nu}$, order parameter $M \sim L^{-\beta/\nu}$, susceptibility $\chi \sim \langle M^2 \rangle \sim L^{\gamma/\nu}$, and n-th logarithmic derivative of the order parameter $V_n = \langle M^n E \rangle / \langle M^n \rangle - \langle E \rangle \sim L^{1/\nu}$. The critical temperature was estimated by the cumulant-crossing method [18], which involves the intersection of the order-parameter cumulant $U_M(T) = 1 - \frac{1}{3}\langle M^4 \rangle / \langle M^2 \rangle^2$ at lattice sizes L and L'. The intersection temperature T_c^* scales as $ln^{-1}(b)$, where $b = L'/L$, giving the extrapolation $T_c(L \to \infty)$ as $b \to 0$. Apart from the expected volume-dependent FSS of the thermodynamic quantities at a first-order transition, this possibility was also examined by consideration of the energy cumulant $U_E(T_c) = 1 - \frac{1}{3}\langle E^4 \rangle / \langle E^2 \rangle^2$, which achieves the asymptotic value of

$U_E^* = \frac{2}{3}$ for a continuous transition, but a value of less than $\frac{2}{3}$ at a first-order transition.

Many runs were made at each lattice size so that a crude estimation of errors was possible by simply taking the standard deviation. (The true errors resulting from histogram analyses are difficult to assess [19].) The number of MC steps (MCS), as well as the number of runs, was increased with L in an effort to overcome the effects of critical-slowing down: The actual numbers used were roughly determined so as to reduce the error bars to an acceptable level.

3.1. Ising Model

Simulations on the Ising model with NN AF in-plane interactions $J_\perp = 1$ and NN F coupling $J_\parallel = -1$ along the c axis were made on lattices $L = 12-33$ using runs from 5×10^5 MCS for the smaller lattices, to 1.2×10^6 MCS for the larger lattices, after discarding the initial 1×10^5 - 2×10^5 MCS for thermalization [17]. Averaging was then made using from 6 runs for the smaller lattices to 15 runs at $L = 30$ and 13 runs at $L = 33$. This gives a respectable 1.56×10^7 MCS used for averaging at the largest lattice size. Histograms were generated at our previous estimate of the critical temperature, $T_H \simeq 2.93$ [20].

The results of applying the cumulant-crossing method to estimate the critical temperature are shown in Fig. 2. Linear scaling in the large-L region gives $T_c = 2.9298(10)$, close to the value of T_H used to generate the data. Good evidence that the transition is continuous is found in our extrapolated value $U_E^* = 0.666678(30)$. FSS of the specific heat at T_c is depicted in Fig. 3. Errors were too large to extract a reasonable value for α/ν and the assumption of XY

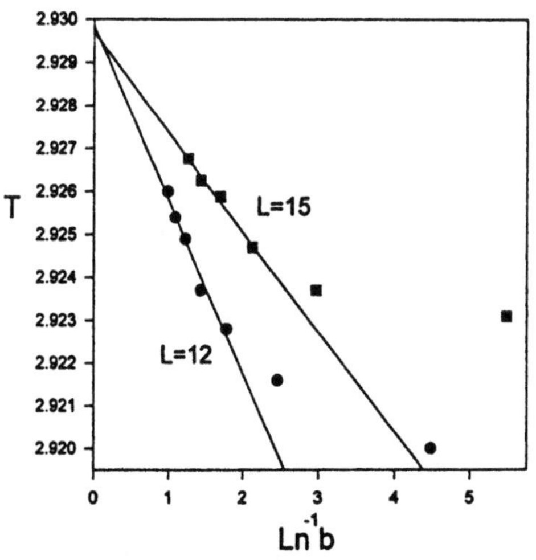

Figure 2: Finite-size scaling of the cumulant-crossing, where $b = L'/L$, for the frustrated Ising model.

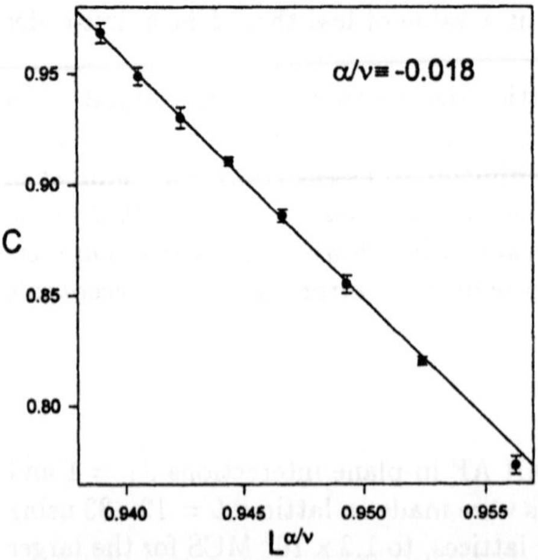

Figure 3: Finite-size scaling of the specific heat for the frustrated Ising model with the assumption of XY universality.

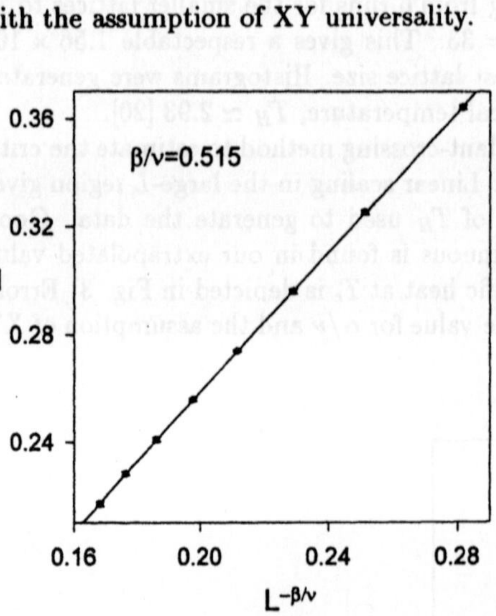

Figure 4: Finite-size scaling of the order parameter for the frustrated Ising model.

universality was made in this case: The results appear to be consistent with this prediction for the larger values of L. Good straight-line fits were found for the order parameter, as shown in Fig. 4, as well as the susceptibility. Stronger finite-size effects appeared when the scaling of V_1 and V_2 was attempted, as summarized in Ref.[20].

Our final estimates for the critical exponents are given in Table 1 along with those from the standard unfrustrated XY model, as well as from Bunker *et al.*, who performed simulations on the same frustrated Ising system as considered

Table 1. Comparison of exponents for the unfrustrated XY model from RG, high-temperature (HT) series expansion, and MC simulations, with those of MC simulations for the Ising model on a stacked triangular lattice (STL). The exponent α is determined from scaling relations.

	α	β	γ	ν
XY:RG $(4-\epsilon)$ [14]	-0.013(15)	0.349(4)	1.315(7)	0.671(5)
XY:HT Series [15]	-0.01(3)	0.348(15)	1.315(9)	0.67(1)
XY:MC [16]	+0.014(20)	0.331(10)	1.324(1)	0.662(7)
STL Ising: our work [17]	+0.012(30)	0.341(4)	1.31(3)	0.662(9)
STL Ising: Bunker *et al.* [8]	-0.05(3)	0.311(4)	1.43(3)	0.685(3)

here. Errors were estimated based on the uncertainty in our value of T_c (i.e., scaling was also made at $T = 2.9288$ and $T = 2.9308$). Our results are fully consistent with the proposal of Berker *et al.* that the frustrated Ising model exhibits XY universality. Although the results of Bunker *et al.* are not too different, they suggest the possibility of a new universality class. We believe this to be a consequence of an incorrect estimation of $T_c \simeq 2.920(5)$ as well as the fact that their results are based on scaling of the extrema of thermodynamic functions (except M) rather than at the critical temperature. Finite-size effects appear to be more pronounced in the former case.

Earlier experimental results for the critical exponents on the quasi-one-dimensional AF's $CsCoCl_3$ and $CsCoBr_3$ are summarized in Ref. [6]. Values for β range from 0.22(2) to 0.35(2) and recent results for α give -0.025(4). New, and highly accurate, neutron diffraction data on $CsCoBr_3$ yield the estimates $\beta = 0.28(2)$, $\gamma = 1.05(8)$, and $\nu = 0.54(5)$ [21]. Except for α, which is consistent with XY criticality, these latter results are suggestive of mean-field tricritical behavior and will be discussed further in Sect. 4.

3.2. XY Model

MC simulations were performed on the XY model with the same quality of statistics as in the Ising case [22]. With vector spins, however, one should also consider the chirality order defined by $\kappa \sim \sum_P \mathbf{S}_i \times \mathbf{S}_j$, where the sum is over a triangular plaquette. Histograms were generated at Kawamura's estimate of the critical temperature $T_H = 1.458$ [11]. For this system, an estimate of the correlation time for the spin order parameter was found to be about 620 MCS at T_H and $L = 24$. For our largest lattice, this implies averaging was made over approximately 500 independent configurations. This compares with only about 8 independent configurations in the MC simulations of Kawamura at the largest lattice used in his work, $L = 60$. We note that Kawamura did not generate histograms and his analysis of exponents was by the standard "data collapsing", method which involves the simultaneous estimation of the critical temperature and critical exponents.

Fig. 5 shows the results of the cumulant-crossing method for the spin and chiral ordering temperatures. As discussed in Ref. [22], the coincidence or

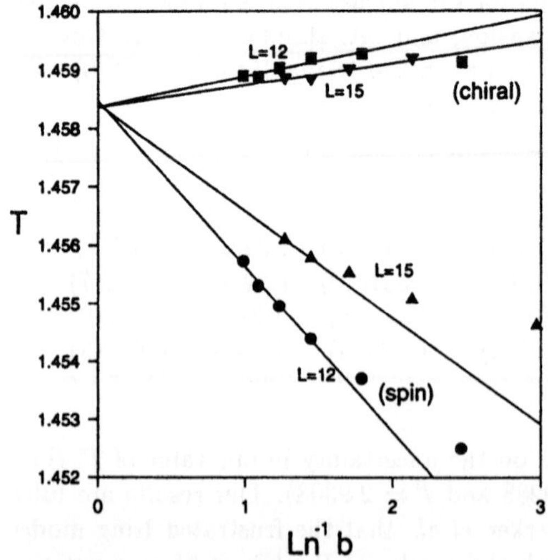

Figure 5: Finite-size scaling of the cumulant-crossing, where $b = L'/L$, for the frustrated XY model.

not of these two critical temperatures has been examined extensively in the literature, with no firm conclusions. Within error, our results suggest that they are the same (or very close), $T_c = 1.4584(6)$. Our estimate $U_E^* = 0.666652(20)$, along with the failure of the assumption of volume-dependent scaling to yield good straight-line fits, indicates that the transition is continuous.

Our final estimates for the critical exponents are given in Table 2, along with those of Kawamura, where our errors again were determined by the uncertainty in T_c. The results are not so different from those of Kawamura, but strongly suggest that the spin-order transition is tricritical (see Sect. 4) and not a new universality class. Our results for the chiral order are more difficult to interpret but the fact that $\nu_\kappa \neq \nu$ suggests that it is somehow decoupled from the spin ordering.

There is a wide variety of experimental systems available to test the various predictions concerning the critical behavior of the frustrated XY model, as summarized in Ref. [6]. Although there is no clear conclusion, more recent results on $CsMnBr_3$ strongly suggest unusual exponents. For example, values for α range from 0.39(9) to 0.40(5), $\beta = 0.21(2) - 0.25(1)$, $\gamma = 1.01(8) - 1.10(5)$, and $\nu = 0.54(3) - 0.57(3)$ in this material. In the case of the rare-earth helimagnets, in addition to evidence suggesting a first-order transition, reported exponent values are $\alpha = 0.10(2) - 0.27(2)$, $\beta = 0.34(1) - 0.39(4)$, $\gamma = 1.05(7) - 1.14(10)$, and $\nu = 0.57(5)$. Corresponding chiral exponents have not been measured [33].

3.3. Heisenberg Model

Although our simulations of the Heisenberg model were made with considerably fewer MCS (about a factor of ten) than for the Ising and XY cases, the quality

Table 2. Comparison of exponents for the frustrated XY model with those expected of tricirital behavior for both spin and chiral (subscript κ) order.

	α	β	γ	ν	β_κ	γ_κ	ν_κ
Kawamura [11]	0.34(6)	0.25(1)	1.13(5)	0.54(2)	0.45(2)	0.77(5)	0.55(2)
Our work [22]	0.46(10)	0.24(2)	1.03(4)	0.50(1)	0.38(2)	0.90(9)	0.55(1)
Tricritical	$\frac{1}{2}$	$\frac{1}{4}$	1	$\frac{1}{2}$			

Table 3. Comparison of exponents for the frustrated Heisenberg model from MC simulations for both spin and chiral (subscript κ) order. Except in the case of Kawamura (first row), all other values given for α were determined by the Rushbrooke equality.

	α	β	γ	ν	β_κ	γ_κ	ν_κ
Ref. [11]	0.24(8)	0.30(2)	1.17(7)	0.59(2)	0.55(4)	0.72(8)	0.60(3)
Ref. [23]	0.25(2)	0.29(1)	1.185(3)	0.586(8)	0.50(2)	0.82(4)	0.61(1)
Ref. [25]	0.25(3)	0.29(1)	1.18(2)	0.585(9)			
Ref. [26]	0.19(4)	0.28(2)	1.25(3)	0.59(1)			

of the FSS data was found to be quite good [23]. In this case, lattices $L = 12 - 36$ were used, with 1×10^6 to 2.6×10^6 MCS for averaging. Both spin and chiral ordering temperatures were found to be the same within error, 0.9577(2) and 0.9577(6), respectively. The estimate $U_E^* = 0.66666(1)$ indicates that the transition is indeed continuous. Our results for the exponents, along with those of Kawamura and two other MC-histogram simulation results, are provided in Table 3. There is general agreement and these four independent sets of data strongly support the proposal of a new $n = 3$ chiral universality class. The only reported experimental candidates to test these predictions are VBr_2 and VCl_2 [24], where values $\alpha = 0.28(5) - 0.59(5)$, $\beta = 0.20(2)$, and $\nu = 0.62(5)$ are not conclusive.

4. Field-Induced 3-State Potts Transitions

A variety of complex phase diagrams and multicritical-point phenomena can result from the application of a magnetic field to these frustrated systems. Many examples are reviewed in Refs. [4], [5], and [6] in case of NN AF interactions along the c axis, for which there are a number of experimental realizations. Much less studied, but possibly of more interest, are systems with NN F c-axis interactions. This is due to the presence of a third-order term in the free energy of the form

$$F_3 \sim (\mathbf{m} \cdot \mathbf{S})(\mathbf{S} \cdot \mathbf{S}) + c.c.$$

where $\mathbf{m}$ is the magnetization induced by the field. Such a term exists only if $3\mathbf{Q} = \mathbf{G}$, and is thus not present in the case of NN AF c-axis interactions where $6\mathbf{Q} = \mathbf{G}$. It gives rise to magnetic order (at finite field) which has the 3-state Potts symmetry. Transitions from the paramagnetic state to such a phase are thus expected to be very weakly first order in 3D.

The magnetic phase diagram of the corresponding Ising model has recently been examined by mean-field theory and histogram-MC simulations, with FSS of thermodynamic functions as described above [27]. Our results indicate a weak first-order transition to the 3-state Potts state for fields larger than about $H = 1$. At smaller fields, however, the transition from the paramagnetic phase is to a state with a different symmetry and we find evidence (in the critical exponents) for a *line* of continuous transitions with tricritical behavior for $0 < H \lesssim 1$. We believe these results may be relevant to explain the anomalous experimental data on the critical exponents of $CsCoBr_3$ at the Néel point, as described in Sect. 3.1. We note that although this material has strong AF c-axis interactions, an identical phase diagram is expected in this case if a *staggered* field (along the c axis) is applied. Coupling between the well developed short-range c-axis order could then generate an effective cubic term of the type given above, driving the system tricritical. (This type of argument was originally applied to the frustrated Ising AF by Heinonen and Petschek [32].)

In the case of the XY model, the phase diagram exhibits a weak first-order transition to the 3-state Potts phase for $0 < H \lesssim 2.5$. [28]. We emphasize here that the $H = 0$ transition is not a usual tricritical point in the mean-field sense since the first-order transition is controlled by a cubic term (linear in H); The fourth-order term remains positive at all fields. For $H < 0$, the cubic term is again non-zero and the transition is first order. $H = 0$ thus represents an isolated point along the paramagnetic boundary in the phase diagram at which a continuous transition occurs. It is nevertheless tempting to attribute the tricritical exponents found in our MC simulations at $H = 0$ to effects of fluctuations involving the 3-state Potts state. Again, an identical phase diagram is found in the case of NN AF c-axis interactions with the application of a staggered field so that there may be some relevance to materials such as $CsMnBr_3$. We note that similar phase diagrams are found for the 2D XY as well as Heisenberg systems [29].

Our continuing interest in the 3-state Potts transition is assured by the suggestion of Berker [30] that a small amount of disorder can drive such weak first-order transitions to be continuous, thus providing the possibility of an effective 3-state Potts universality class in 3D. For this purpose, we intend to examine by FSS the tranition in XY model with NN F interactions at $H = 1.5$. Preliminary results are reported in Ref. [28], where FSS of limited MC-histogram data on the extrema in various thermodymnamic functions was performed in order to demonstrate the weak first-order character of the transition as well as make a crude estimate of the latent heat. Since the latter quantity will be used as the primary indicator of when the transition becomes continuous with the addition of disorder, it is desirable to check the accuracy of these earlier results [31]. Much more extensive data (at $H = 1.5$ and no disorder) has now

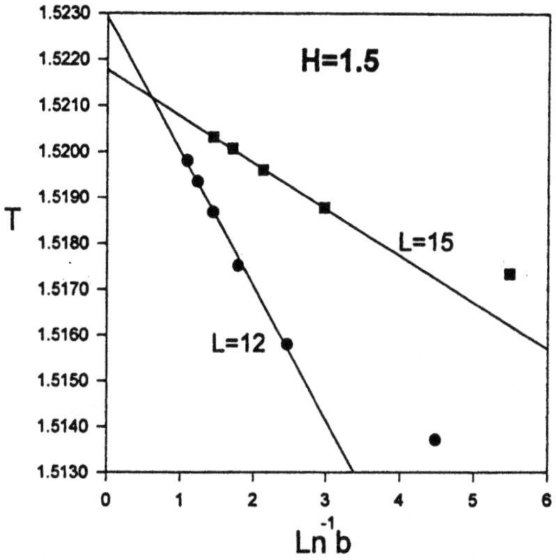

Figure 6: Finite-size scaling of the cumulant-crossing, where $b = L'/L$, for the XY model in a field.

been generated and the results of FSS of these new resutls *at* the critical point are discussed below.

Simulations were performed at the previous estimate of the critical temperature $T_H = 1.522$ on lattices $L = 12 - 30$ using from 1×10^6 to 1.8×10^6 MCS per run for averaging. Final averages were calculated using 6-12 runs for L = 12-27, and 3 runs at L =30. This represents about a factor of 10 more MCS than was used in Ref. [28], although $L = 33$ was also included in the earlier runs. Results of the cumulant-crossing method shown in Fig. 6 give the estimate $T_c = 1.5225(10)$. Scaling of the specific heat and energy cumulant with volume are shown in Fig. 7, where error bars on the $L = 30$ data are not included since only 3 runs were made. Notable is that volume-dependent scaling occurs in these quantities only at the larger values of L. This behavior was also observed in the earlier results on the scaling of the extrema of these functions. The estimate $U_E^* = 0.66642(3)$ compares well with the previous result $U_E^* = 0.66643(3)$. Estimation of the latent heat both from the slope of the specific heat, as well as the value of U_E^*, gives $E_{T_N^+} - E_{T_N^-} \simeq 0.034$ in agreement with the previous results of 0.032. This can be compared with the value 0.222 for the 3-state Potts model in 3D (see Ref. [28]), considered to be very weakly first order.

Scaling of the suceptibility defined by $\chi' \sim \langle M^2 \rangle - \langle M \rangle^2$, as well as V_1 are shown in Fig. 8. Near perfect volume dependence is observed in the five largest lattice sizes. No clear volume dependence was found for the susceptibility defined by $\chi \sim \langle M^2 \rangle$, presumably due to the discontinuity in the order parameter. This latter definiton thus appears to be useful only at a continuous transition.

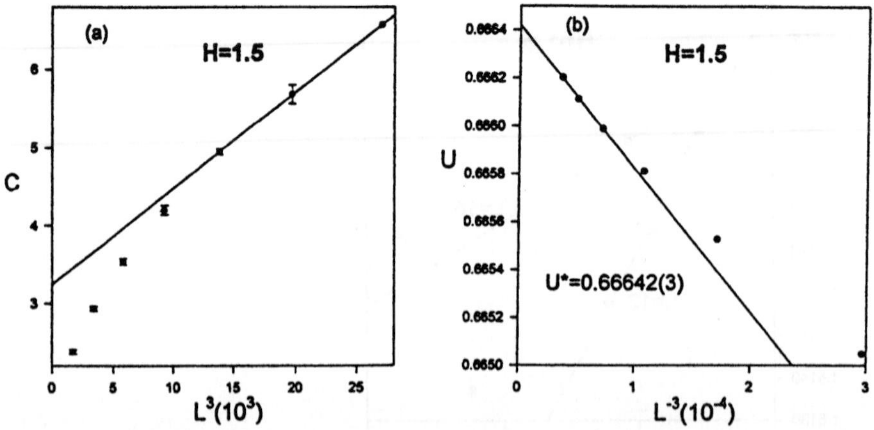

Figure 7: Finite-size scaling of the (a) specific heat and (b) energy cumulant (where L=12 has been omitted for clarity) for the XY model in a field. Straight-line fits are through the three largest lattice sizes only.

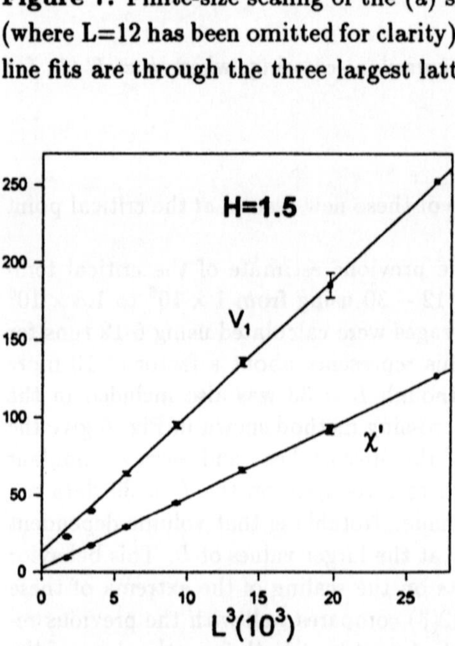

Figure 8: Finite-size scaling of the susceptibility and logarithamic derivative of the order parameter for the XY model in a field. Straight-line fits are through the five largest lattice sizes only.

5. Conclusions

Several conclusions can be made concerning MC simulations of phase transitions in frustrated systems. The critical exponents obtained from finite-size scaling of histogram-generated data for the Ising model are in agreement with the symmetry arguments of Berker *et al* which suggest XY universality; for the XY model, they support the posibility raised by Azaria *et al.* of tricritical behavior (although their formalism is strictly applicable for $n \geq 3$; and Kawamura's proposal of a new $n = 3$ chiral universality class is fully supported by our results (as well as others) for the Heisenberg model. We expect, however, that this is not the final word regarding the critical behavior of these systems.

It has also been emphasized in this work that a magnetic-field induced weak first-order transition occurs with NN F c-axis interactions. Such behavior can be observed in volume-dependent scaling of thermodynamic functions, although those which involve energy fluctuations require larger lattice sizes. It appears that reasonable results (e.g., for the latent heat) can be obtained with less computational effort than in the cases of a continuous transition.

Acknowledgements

We thank P. Azaria, A.N. Berker, A. Chubukov, M.F. Collins, B. Delamotte, H.T. Diep, A. Ferrenberg, B. Gaulin, O. Heinonen, H. Kawamura, S. Miyashita, G. Zumbach, and especially A. Caillé for useful discussions, as well as the Service de l'Informatique for partial use of 10 RISC 6000 Workstations. This work was supported by NSERC of Canada and FCAR du Québec.

References

[1] P. Azaria and B. Delamotte, in *Magnetics Systems with Competing Interactions*, ed. by H.T. Diep (World Scientific, Singapore, 1994).

[2] A.M. Ferrenberg and R.H. Swendsen, Phys. Rev. Lett. **61**, 2635 (1988); *ibid* **63**, 1195 (1989); Computers in Physics, Sept./Oct., 101 (1989); P. Peczak, A.M. Ferrenberg, and D.P. Landau, Phys. Rev. B **43**, 6087 (1991); A.M. Ferrenberg and D.P. Landau, Phys. Rev. B **44**, 5081 (1991); K. Chen, A.M. Ferrenberg, and D.P. Landau, Phys. Rev. B **48**, 3249 (1993).

[3] P.W. Leung and C.L. Henley, Phys. Rev. B **43**, 752 (1991); P.D. Coddington and L. Han, Phys. Rev. B **50**, 3058 (1994).

[4] M.L. Plumer and A. Caillé, J. Appl. Phys. **70**, 5961 (1991).

[5] M.L. Plumer, A. Caillé, A. Mailhot, and H.T. Diep (see Ref. [1]).

[6] B.D. Gaulin (see Ref. [1]); O.A. Petrenko and M.F. Collins, Can. J. Phys. (to be published).

[7] A.N. Berker *et al.*, J. Appl. Phys. **55**, 2416 (1984); D. Blankschtein *et al.*, Phys. Rev. B **29**, 5250 (1984).

[8] A. Bunker, B.D. Gaulin, and C. Kallin, Phys. Rev. B **48**, 15861 (1993).

[9] D.H. Lee, J.D. Joannopoulos, J.W. Negele, and D.P. Landau, Phys. Rev. Lett. **52**, 433 (1984); Phys. Rev. B **33**, 450 (1986).

[10] H. Kawamura and S. Miyashita, J. Phys. Soc. Japan **53**, 4138 (1984).

[11] H. Kawamura, J. Phys. Soc. Japan **54**, 3220 (1985); **55**, 2095 (1986); **56**, 474 (1987); **61**, 1299 (1992); in *Recent Advances in Magnetism of Transition*

Metal Compounds, ed. by A. Kotani and N. Suzki (World Scientific, Singapore, 1993).
[12] G. Zumbach, Phys. Rev. Lett. **71**, 2421 (1993); Nucl. Phys. B **413**, 754, 771 (1994).
[13] S.A. Antonenko and A.I. Sokolov, Phys. Rev. B **49**, 15901 (1994).
[14] J.C. Le Guillou and J. Zinn-Justin, J. Phys. Lett. (Paris) **46**, L137 (1985).
[15] P. Butera, M. Comi, and A.J. Guttmann, Phys. Rev. B **48**, 13987 (1993).
[16] A.P. Gottlob and M. Hasenbusch, Physica A **210**, 593 (1993).
[17] M.L. Plumer and A. Mailhot, Phys. Rev. B (to be published as a *Comment* on Ref.[8]).
[18] K. Binder, Z. Phys. B **43**, 119 (1981).
[19] A.M. Ferrenberg, D.P. Landau, and K. Binder, J. Stat. Phys. **63**, 867 (1991).
[20] M.L. Plumer *et al.*, Phys. Rev. B **47**, 14312 (1993).
[21] M. Mao, R.B. Rogge, and B.D. Gaulin (private communication).
[22] M.L. Plumer and A. Mailhot, Phys. Rev. B **50**, 16113 (1994).
[23] A. Mailhot, M.L. Plumer, and A. Caillé, Phys. Rev. B **50**, 6854 (1994).
[24] K. Takeda *et al.*, J. Phys. Soc. Japan **55**, 727 (1986); H. Kadowaki *et al.*, J. Phys. Soc. Japan **56**, 4027 (1987); J. Wosnitza *et al.*, J. Phys.: Condens. Matter **6**, 8045 (1994).
[25] T. Bhattacharya *et al.*, J. Phys. I (France) **4**, 122 (1994).
[26] D. Loison and H.T. Diep (unpublished).
[27] M.L. Plumer and A. Mailhot, Phys. Rev. B (to be published).
[28] M.L. Plumer, A. Mailhot, and A. Caillé, Phys. Rev. B **48**, 3840, 15835 (1993).
[29] D.H. Lee, J.D. Joannopoulos, J.W. Negle, and D.P. Landau, Phys. Rev. B **33**, 450 (1986); H. Kawamura and S. Miyashita, J. Phys. Soc. Japan **54**, 4530 (1985).
[30] A.N. Berker, J. Appl. Phys. **70**, 5941 (1991).
[31] R. Fisch, Phys. Rev. B **46**, 11310 (1992).
[32] O. Heinonen and R.G. Petschek, Phys. Rev. B **40**, 9052 (1989).
[33] M.L. Plumer, H. Kawamura, and A. Caillé, Phys. Rev. B **43**, 13786 (1991).

Simulation of Ionic Surfaces from an Absolutely Convergent Solution of the Madelung Problem

D. Wolf

Materials Science Division, Argonne National Laboratory,
Argonne, IL 60439, USA

Abstract. The classic Madelung problem is cast into an absolutely convergent form that is readily evaluated by direct lattice summation, revealing a net r^{-5} range of the net Coulomb potential in ionic crystals and liquids. The realization that Coulomb interactions in condensed systems can actually be rather short ranged (provided the system is overall neutral) leads to the prediction, verified by computer simulations for rocksalt-structured surfaces, that all surfaces in predominantly ionic crystals should be fundamentally reconstructed. The work also provides a conceptual framework for the theoretical treatment of polar surfaces, as demonstrated for the case of the (111) surfaces of NaCl and MgO.

1. Introduction

The classic Madelung problem, i.e., the divergence associated with the r^{-1} term in the Coulomb potential of condensed systems [1], and its consequences for the physics of ionic crystals and liquids have received considerable attention throughout this century. The mathematical problems associated with the handling of conditionally convergent series have led to computationally expensive - and physically not very transparent - summation methods which, based mostly on Ewald's solution [2], are now in common use for the simulation of ionic materials. These problems have also given rise to a widespread belief that certain "typically ionic" phenomena, such as the divergence of the energies of polar surfaces or the long-range charge ordering in ionic liquids, are a consequence of the long-ranged Coulomb interactions. [3] However, as evidenced, for example, by Evjen's solution of the Madelung problem [4] and by extensive simulations of ionic liquids [3], in many instances Coulombic effects seem to cancel almost completely at long range.

The convergence problems encountered when simply summing the Coulomb potential, $\pm r^{-1}$, over shells of the rocksalt lattice are illustrated in Fig. 1. Figure 1(a) (left half) reveals large oscillations of the apparent Madelung energy thus obtained between large positive and negative values; its value depends strongly on the terminating crystal shell, with no indication of convergence towards the dashed line which gives the correct value, $E_{Mad}=-3.495116a^{-1}$ (where a is the cubic lattice parameter). Figure 1(b) (right half), showing the apparent Madelung energy as a function of the difference between the total number of cations and anions up to a given shell, N(+)-N(-), gives some indication as to the origin of this behavior since

Springer Proceedings in Physics, Vol. 80
Computer Simulation Studies in Condensed-Matter Physics VIII
Editors: D.P. Landau · K.K. Mon · H.-B. Schüttler

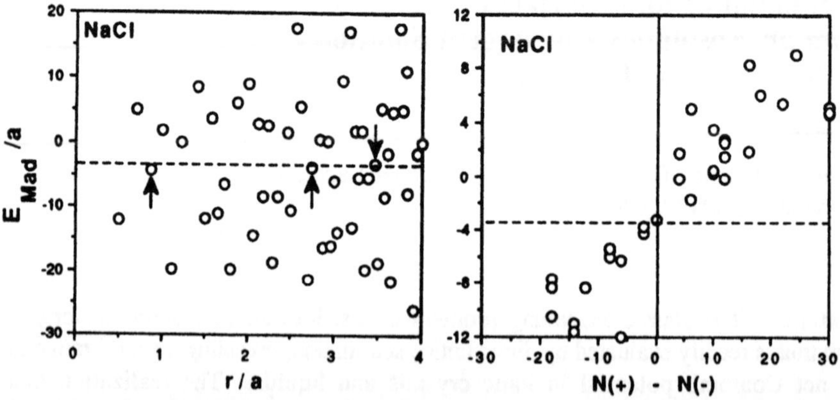

Fig. 1. Total apparent Madelung energy, E_{Mad} (in units of the inverse lattice parameter, a^{-1}) obtained by simply summing the Coulomb potential, $\pm r^{-1}$, over shells of the rocksalt lattice. (a) (left half): E_{Mad} against shell radius, r (in units of the cubic lattice parameter, a); (b) (right half): E_{Mad} against the difference, N(+)-N(-), between the total number of cations and anions up to a given shell of radius r.

N(+)-N(-) is proportional to the net charge of the crystal up to the distance r. In fact, when the system is neutral or almost neutral, values close to the correct Madelung energy are obtained. Three such data points are marked by the arrows in Fig. 1(a), corresponding to the following shell radii and values of N(+)-N(-) (in parentheses): 0.866a (-2), 2.739a (-2) and 3.464a (0). These observations suggest that the large oscillations of the apparent Madelung energy arise directly from the fact that the crystal lattice is practically never neutral when terminated by complete *crystal* shells.

Based on the above observation, in the present work a simple, absolutely convergent solution of the Madelung problem involving direct r^{-1} lattice summation, is presented. (For a preliminary report, see Ref. [5].) As illustrated in Sec. 2, operationally the "trick" is to sum over the *neutral* shells of the *Bravais* lattice and not, as in Fig. 1, over the *charged* shells of the *crystal* lattice. The convergence analysis presented in Sec. 3 demonstrates that the effective Coulomb potential in ionic crystals and liquids decreases as r^{-5}. This realization leads naturally to the prediction, illustrated here for materials with NaCl structure, that most ionic-crystal surfaces should reconstruct in a manner that involves "molecular" building blocks (see Sec. 4). The insights gained from understanding the nature of this convergence also provide a conceptual framework for the theoretical treatment of polar surfaces, another long-standing problem in the physics of ionic materials (see Sec. 4).

2. A Dipolar Model

As illustrated above, the main difficulty in the evaluation of Madelung's constant by direct lattice summation arises from the fact that most shells of the *crystal* lattice are charged and that, therefore, it is virtually impossible to terminate the summation in a

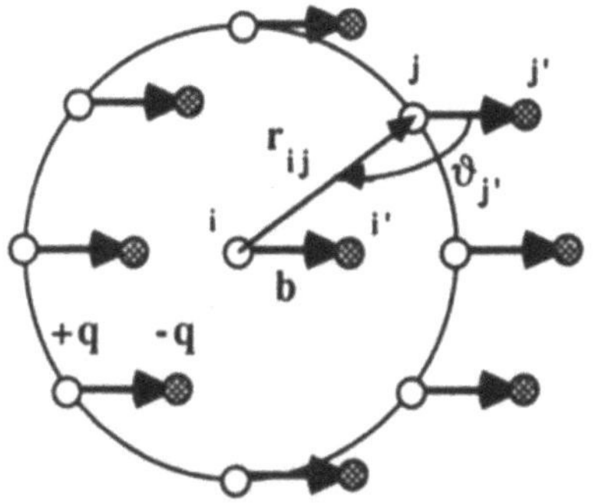

Fig. 2. Neutral shells of NaCl "molecules" are obtained by attachment of the NaCl basis dipoles to the sites of the fcc *Bravais* lattice (schematic). $r_s \equiv r_{ij}$ denotes the radii of the shells of the Bravais lattice.

way that renders the system as a whole neutral. Figure 2 demonstrates how operationally this problem may be overcome by simply summing over *neutral* shells of basis molecules, i.e., shells of the *Bravais* lattice with subsequent attachment of the neutral basis molecule (such as NaCl, with charges $\pm q$). This results in the generation of two identical, oppositely charged sublattices displaced relative to each other by the basis vector **b**. The total "molecular" Coulomb energy, E^{tot}_{mol}, of some ion i at the origin is then given by [5]

$$E^{tot}_{mol} = -q^2/b + q^2 \Sigma_{j \neq i} \Sigma_{j'} (1/r_{ij} - 1/r_{ij'}) = E_{intra} + \Sigma_{r_s} E_{inter}(r_s) \ , \qquad (1)$$

where the first term represents the "intra-molecular" (i-i') interaction while the second is the "inter-molecular" interaction of ion i with the molecules in shells with radii $r_{ij} \equiv r_s$ (see Fig. 2).

Intuitively one would expect the double sum in Eq. (1) to converge rapidly for the following two reasons. First, because the direction of **b** is fixed while that of $\mathbf{r}_{ij}$ is averaged over a discrete set of Bravais points on a sphere, within a given Bravais-lattice shell (i.e., for a fixed value of $r_{ij} \equiv r_s$), the values of $1/r_{ij} - 1/r_{ij'}$ vary between small positive and negative (for $\vartheta_{j'} > 90°$ and $\vartheta_{j'} < 90°$, respectively; see Fig. 2). By contrast with a sum over *charged* shells of the *crystal* lattice (involving very large, non-compensated $\pm 1/r_{ij}$ terms), the sum in Eq. (1) over Bravais shells therefore involves differences between very small terms. Second, these already small positive or negative net values obtained for each shell are summed over the entire Bravais lattice, with rapidly decreasing small negative and positive values as r_s increases.

The rapid convergence of the total sum, E^{tot}_{mol}, in Eq. (1) is illustrated in Fig. 3 for an fcc Bravais lattice with a dipolar NaCl basis (see also Fig. 2). Notice that the contributions from complete dipolar shells in Eq. (1), $E_{inter}(r_s)$ (right scale in Fig. 2), indeed fluctuate about zero while decreasing rapidly with increasing shell radii.

An apparent problem with the dipolar sum in Fig. 3 is that it does not converge to the correct Madelung energy, $E_{Mad} = -3.49513\ q^2/a$; instead, the value $E_{dip} = -1.40074\ q^2/a$ is obtained. However, a sum over shells of dipoles *cannot* be

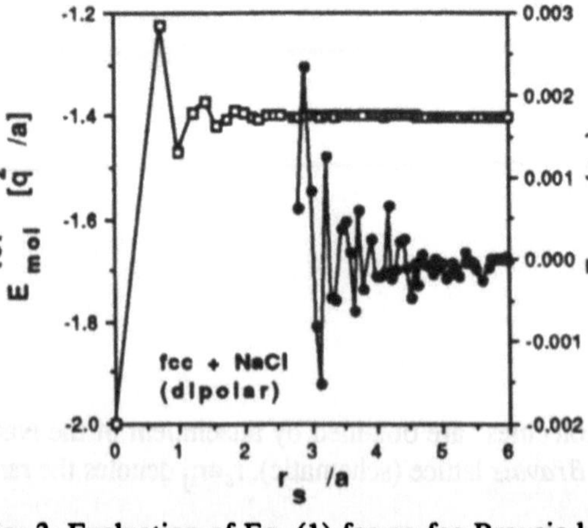

Fig. 3. Evaluation of Eq. (1) for an fcc Bravais lattice with dipolar NaCl basis. [5] Filled circles (right scale): shell-by-shell contributions to the energy in Eq. (1). Open squares (left scale): sum of all shell contributions. The point at $r_s=0$ corresponds to the intra-molecular energy, -1/0.5a = -2/a.

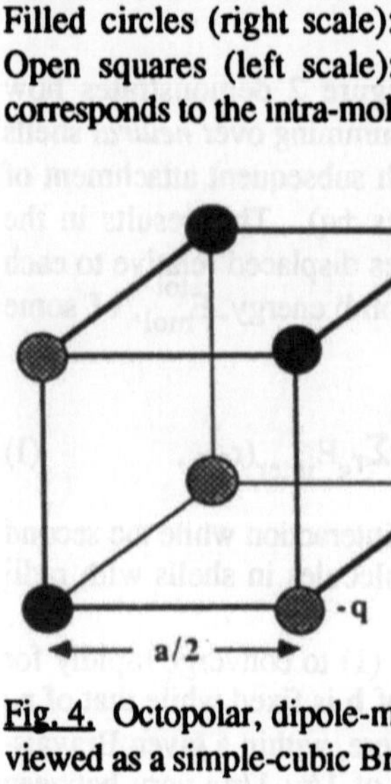

Fig. 4. Octopolar, dipole-moment-free $(NaCl)_4$ building block of the rocksalt lattice viewed as a simple-cubic Bravais lattice. a is the cubic lattice parameter.

terminated without rendering the system as a whole polarized (see Fig. 2). To obtain the correct Madelung energy, the polarization energy, E_{pol}, has to be subtracted; i.e.,

$$E_{Mad} = E_{mol}^{tot} - E_{pol} . \tag{2}$$

Since each molecule contributes a dipole moment of qb to the polarization (Fig. 2), the related dipole energy per unit volume is given by [6]

$$E_{pol} = (2\pi/3\Omega)\,(qb)^2, \tag{3}$$

where Ω is the "molecular" volume. In the fcc lattice, $\Omega=a^3/4$; with $|b|=a/2$ for the

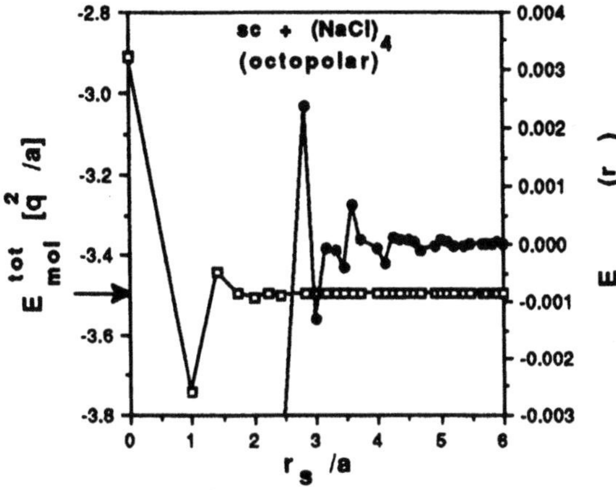

Fig. 5. Evaluation of Eq. (1) for an sc Bravais lattice with the octopolar $(NaCl)_4$ basis in Fig. 4. [5] Filled circles (right scale): shell-by-shell contributions in Eq. (1) against shell radius, r_s. Open squares (left scale): sum of all shell contributions, including the intramolecular contribution (see also Fig. 3.) The arrow indicates the correct Madelung energy.

NaCl basis molecule, we obtain $E_{pol} = (2\pi/3)\ q^2/a = 2.09439$ which, when subtracted from E^{tot}_{mol} in Fig. 3, gives the correct Madelung energy.

By choosing a basis molecule without a dipole moment, this polarization correction can be avoided altogether. Thus, instead of viewing the NaCl lattice as an fcc Bravais lattice with a *dipolar* basis, one can choose the simple cubic (sc) Bravais lattice with the cube-shaped *octopolar* $(NaCl)_4$ basis shown in Fig. 4. Avoiding thus the generation of a long-range polarization of the system, the direct r^{-1} Coulomb sum based on Eq. (1) (in which, however, dipolar molecules and shells are replaced by octopolar ones) gives the Madelung energy directly, as demonstrated in Fig. 5.

An interesting property of the direct-summation method involving octopolar molecules is that the intra-molecular contribution (see Fig. 4),

$$E_{intra} = (q^2/a)\ [-6 + 3\sqrt{2} - 2/\sqrt{3}] \approx -2.91206\ (q^2/a) \quad , \qquad (4)$$

is identical to that obtained from Evjen's method [4,7] while avoiding the ambiguities associated with the assignment of fractional charges to the ions in the unit cell. The latter is particularly important when considering defected ionic crystals.

The above results are summarized in Fig. 6 showing the Madelung energy obtained by direct summation for rocksalt viewed as (a) an fcc Bravais lattice with dipolar NaCl basis (see Fig. 3 and Eq. (2)) and (b) an sc Bravais lattice with octopolar $(NaCl)_4$ basis. In the latter case, $E_{Mad} \equiv E^{tot}_{mol}$ (see Eqs. (1) and (2), with $E_{pol} \equiv 0$). This comparison demonstrates that the dipolar and octopolar sums converge to identical values. This suggests that the sum over octopoles may be broken down into the contributions due to the four dipoles 1-2 to 7-8 forming the octopole (see

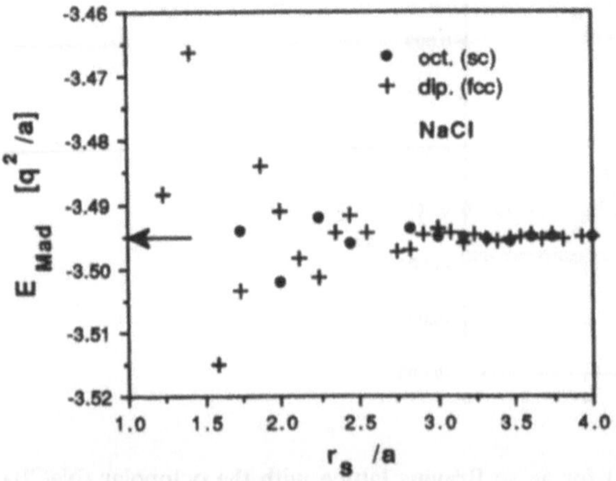

Fig. 6. Madelung energy against cutoff radius obtained by direct lattice summation involving either a dipolar (fcc) basis (Figs. 2 and 3 with Eqs. (2) and (3)) or an octopolar (sc) basis (Figs. 4 and 5). The arrow indicates the correct Madelung energy.

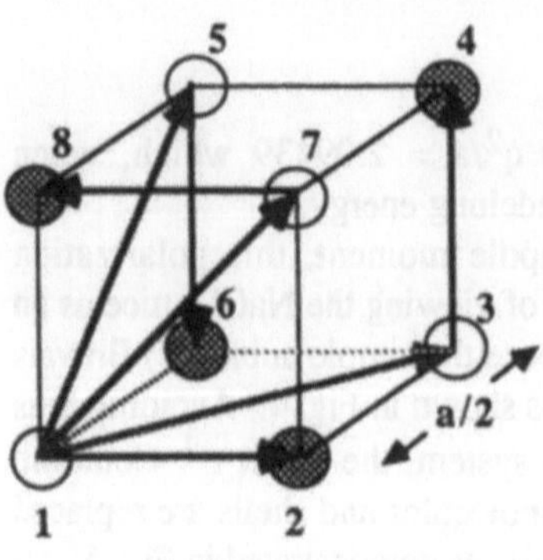

Fig. 7. Breakdown of the sum over octopoles into four contributions from dipoles labeled 1 - 2 to 7 - 8; see also Table 1.

Table 1. Determination of the *octopolar* Madelung energy of the (sc) NaCl lattice as a sum of the potentials of four interpenetrating sc Bravais lattices with the *dipolar* bases 1 - 2 to 7 - 8 sketched in Fig. 7.

dipole	E_{mol}^{tot} [q^2/a]
1 - 2	-2.21775
3 - 4	+0.74293
5 - 6	-1.01015
7 - 8	-1.01015
Total [E_{Mad}]	-3.49512

Fig. 7). As seen from the values listed in Table 1, summing the four dipole potentials at the site of any ion reproduces, indeed, the correct Madelung constant.

That a direct evaluation of the Madelung energy of the NaCl lattice involving octopolar molecules is possible merely with pencil and paper was pointed out earlier by Lacman [8]. His "octopolar approximation" is based on the fact that the interaction between complete octopolar molecules decreases as r_s^{-7} [9]. The above analysis demonstrates that it is not necessary to consider the interactions only between complete octopoles. Instead, while avoiding the polarization problem, this "approximation" follows from the more fundamental r_s^{-5} convergence behavior for the summation involving dipolar shells of the Bravais lattice (see the next section).

3. Convergence

A formal investigation of the convergence behavior of Eq. (1) with Fig. 2 starts with a power expansion of the inter-molecular terms with subsequent summation over the dipolar shells of the Bravais lattice. Introducing the angle $\vartheta_{j'}$ between $\mathbf{r}_{ij}$ and $\mathbf{b}$ (see Fig. 2), these terms may be written in the familiar manner as

$$1/r_{ij} - 1/r_{ij'} = (1/r_{ij})\,\{1 - [1 + (b/r_{ij})^2 - 2(b/r_{ij})\cos\vartheta_{j'}]^{-1/2}\}\ . \tag{5}$$

For $b/r_{ij} \ll 1$, the square root can be expanded into a power series and inserted into Eq. (1). We define the angular sums over complete dipolar Bravais shells by

$$\langle\cos^n\vartheta\rangle_{r_s} = \frac{1}{N(r_s)}\,\Sigma_{j'(r_s)}\cos^n\vartheta_{j'}\ , \tag{6}$$

where $N(r_s)$ is the number of Bravais sites (i.e., NaCl molecules) in a given shell, labeled by $j'(r_s)$. Equation (1) may then be written as follows [5]:

$$E_{mol}^{tot} = -\frac{q^2}{b}\,\{1 + \Sigma_{r_s}N(r_s)\,[(\tfrac{b}{r_s})^2\langle\cos\vartheta\rangle_{r_s} - \tfrac{1}{2}(\tfrac{b}{r_s})^3(1-3\langle\cos^2\vartheta\rangle_{r_s}) - \tfrac{1}{2}(\tfrac{b}{r_s})^4 (3\langle\cos\vartheta\rangle_{r_s} - 5\langle\cos^3\vartheta\rangle_{r_s}) + \tfrac{1}{8}(\tfrac{b}{r_s})^5(3-30\langle\cos^2\vartheta\rangle_{r_s} + 35\langle\cos^4\vartheta\rangle_{r_s}) + ..]\}\ . \tag{7}$$

Because of the inversion symmetry of Bravais lattices, all odd powers of the cosine summed over the dipolar shells vanish identically. With the exception of the quadratic term (for which $\langle\cos^2\vartheta\rangle_{r_s}=1/3$), the even powers in Eq. (6) vary from one shell to another, and no simple r_s-independent values attain. However, using these properties, the leading r_s-dependent term in Eq. (7) is found to be of 5-th order, [9]

$$E_{mol}^{tot} \approx E^{(5)} = \frac{7}{8}\frac{q^2}{b}\,\Sigma_{r_s}N(r_s)\,(\tfrac{b}{r_s})^5(1 - 5\langle\cos^4\vartheta\rangle_{r_s})\ , \tag{8}$$

while the next higher-order terms are $E^{(7)} \sim r_s^{-7}$ and $E^{(9)} \sim r_s^{-9}$, etc. The convergence of the Coulomb energy in Eq. (1) is therefore determined by $E^{(5)}$.

A shell-by-shell comparison of the analytical expression for $E^{(5)}$ in Eq. (8) and the sum of the three lowest-order terms, $E^{(5)}+E^{(7)}+E^{(9)}$, with the results determined by direct evaluation of Eq. (1) is shown in Fig. 8 (for the fcc Bravais lattice with

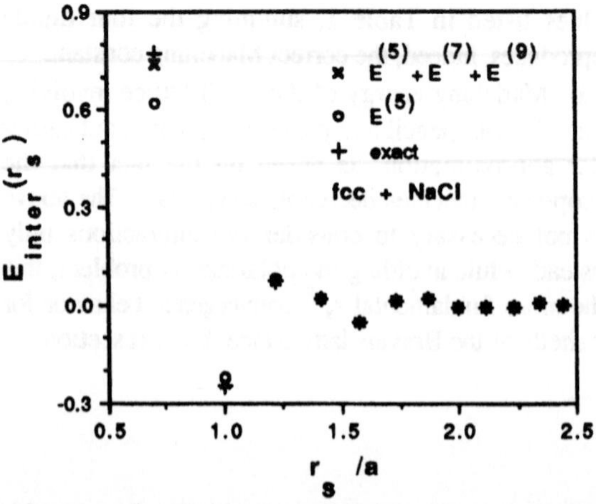

Fig. 8. Coulomb energy for Bravais shells of radii r_s determined analytically from Eq. (8) for $E^{(5)}$, from Eqs. (7) and (8) for the sum of the three lowest-order terms, $E^{(5)}+E^{(7)}+E^{(9)}$, and numerically (i.e., exact results) by directly evaluating Eq. (1) for the fcc Bravais lattice with dipolar basis (see also Fig. 3, right scale).

dipolar basis). According to Fig. 8, for Bravais-shell radii $r_s>a$, the lowest-order analytical results obtained from Eq. (8) agree completely with the direct results.

According to Eq. (8), $E^{(5)}$ vanishes identically for any shell for which $<\cos^4\vartheta>_{r_s}=0.2$, and the signs of any non-vanishing contributions are given by the deviation from this value. As seen from Fig. 9, the shell-by-shell values of $<\cos^4\vartheta>_{r_s}$ scatter practically randomly about 0.2, giving rise to the fluctuations

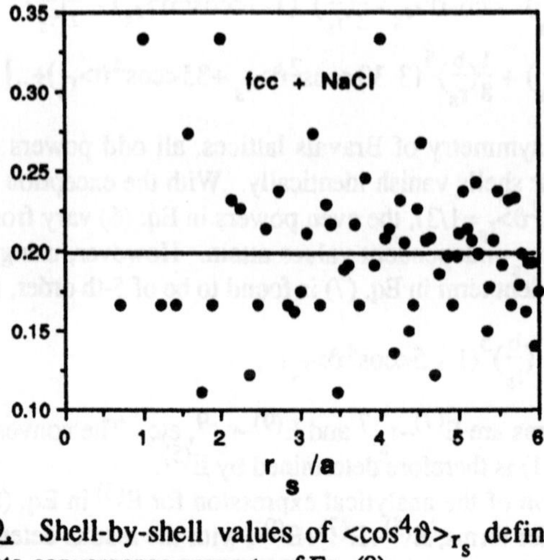

Fig. 9. Shell-by-shell values of $<\cos^4\vartheta>_{r_s}$ defined in Eq. (6), governing the absolute-convergence property of Eq. (8).

about zero in $E_{inter}(r_s)$ in Figs. 3 (right scale) and 8. We finally mention that, since the order in which the individual r^{-1} terms in Eq. (1) are summed does not matter, the direct summation sketched in Fig. 2 and Eq. (1) is absolutely convergent.

That the above arguments are not limited to a perfect crystal lattice is readily seen. For example, Coulomb interactions in liquids are usually evaluated by using 3-d periodic cell borders [2]. By considering the entire simulation cell as the "molecule" attached to the sc lattice of the periodic cell images, the problem is reduced to one analogous to Fig. 2 and Eq. (1). Even in a liquid the r^{-5} convergence therefore attains, an assertion verified by recent molecular-dynamics simulations of liquids.

4. Free Surfaces

The above convergence analysis shows that, from a Coulomb point of view, the rocksalt structure should be viewed as an sc Bravais lattice with an octopolar basis rather than an fcc Bravais lattice with a dipolar basis. This raises the question as to whether free surfaces should not also be made up of octopolar building blocks. [8]

As illustrated in Fig. 10 for the (110) and (111) surfaces in the rocksalt structure, creation of atomically flat surfaces generally requires a breaking up of the octopoles (except for the special case of the (100) surface). Table 2, giving the relaxed zero-temperature energies of the three principal surfaces in NaCl and MgO for both the atomically flat (fcc) surface and its octopolar (sc) reconstruction, illustrates that the breaking of up octopoles, indeed, increases the surface energy substantially.

The effect of octopolar reconstruction is particularly pronounced for the charged (111) surfaces whose *fcc* energy is infinite (see Fig. 11). This divergence is well-known to arise from the long-range dipole moment in the $|A^+C^-B^+A^-C^+B^-|$ stacking of (111) NaCl planes created when attaching NaCl *dipoles* to the |ABC| stacking of *fcc* (111) planes.

By contrast, the $|B^{-1/4}A^{+3/4}C^-B^+A^-C^+B^{-3/4}A^{+1/4}|$ stacking obtained when attaching $(NaCl)_4$ *octopoles* to the |ABC| stacking of *sc* (111) planes (see Fig. 12)

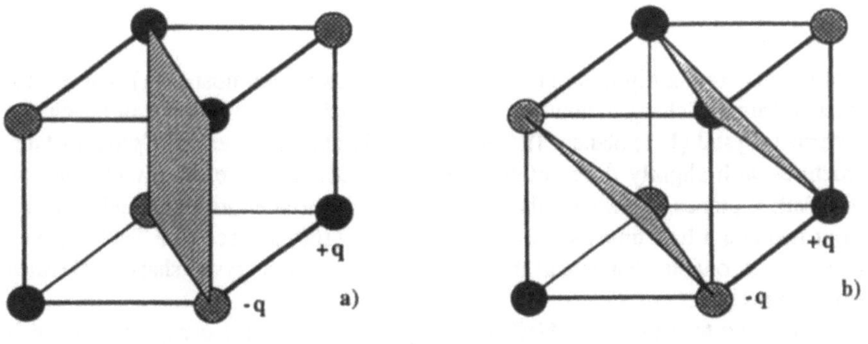

Fig. 10. Creation of atomically flat (a) (110) and (b) (111) surfaces in (fcc) rocksalt requires the breaking up of the octopoles.

Table 2. Relaxed energies (in mJ/m^2) of octopolar and flat surfaces of NaCl [5] and MgO [10] using the "type-2" potentials of Catlow et al. [11] and Catlow et al. [12], respectively. The NaCl results for the (100) and flat (110) surfaces are slightly higher than those given by Tasker [13] due to his use of the shell model, by contrast with the present results. The energies and structures determined via direct r^{-1} lattice summation and by means of Ewald's method agree quantitatively. [5,10]

	NaCl		MgO	
surface	fcc+dip.	sc+oct.	fcc+dip.	sc+oct.
(100)	223	223	1224	1224
(110)	456	366	2490	1993
(111) "+" term.	∞	539	∞	2879
(111) "-" term.	∞	546	∞	2899

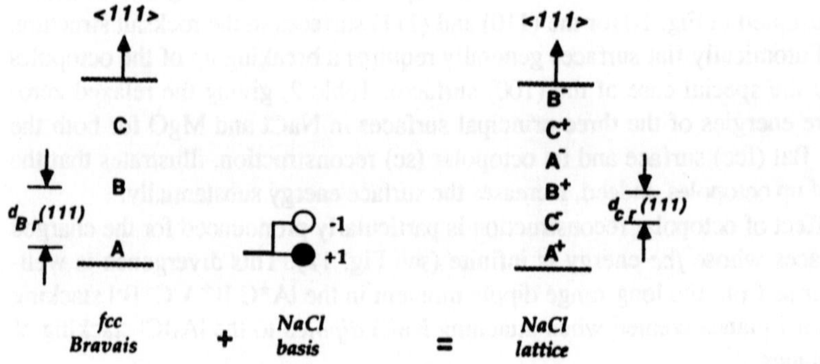

Fig. 11. Creation of the polar structure of the (111) surface in rocksalt by attachment of <111>-oriented NaCl basis molecules to an ..|ABC|.. stack of (111) planes in the fcc Bravais lattice. In accordance with the flat-plate capacitor problem, the energy of the slab diverges for infinitely separated, oppositely charged surfaces.

does not generate a dipole moment, although the two outermost (111) planes thus generated are still charged, however with only 1/4 and 3/4, respectively, of the charge of perfect-crystal (111) planes. The two oppositely charged octopolar "ground-state" structures with slightly different energies thus predicted therefore involve the top *three* lattice planes. However, the energies of both surfaces are finite and typically about two and a half times as large as that of the (100) surface (see Table 2); i.e., there is a reasonable chance that the surface appears in the crystal shape at elevated temperatures [14].

According to Table 2, in NaCl the cation-terminated (1/4 occupied) surface is favored slightly over the anion-terminated (1/4 occupied) surface whereas in MgO the anion-terminated surface has the lower energy. This difference arises from an intricate balance between the Coulomb and short-range-repulsive contributions to the total

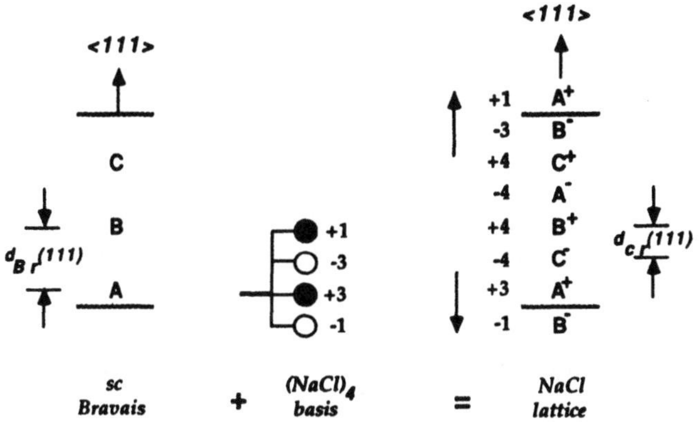

Fig. 12. Creation of the octopolar "ground-state" structure of the (111) surface in rocksalt by attachment of <111>-oriented octopolar molecules (see Fig. 10) to an ..|ABC|.. stack of (111) planes in the sc Bravais lattice. The (either positively or negatively charged) outermost planes contain only 1/4 of the ions of a regular, fully occupied ideal-crystal (111) plane. The next plane towards the bulk is 3/4 filled and oppositely charged while, finally, the third plane from the surface is completely filled. This octopolar reconstruction involving the top three surface planes eliminates the long-range dipole moment of the flat but polar (fcc-based) surface in Fig. 11.

(111)-surface energy, indicating possibly a strong materials dependence of the structure of charged surfaces, by contrast with the structure of neutral surfaces. It also indicates that the structure and energy of charged surfaces are not dominated by Coulombic effects but, instead, by the short-range repulsion between the ions, i.e., by size effects. Unfortunately, no experimental verification of this octopolar reconstruction of either the (110) or the (111) surface is to date available from experiments nor has the net charge of the terminating (111) plane been determined.

We finally mention that the predicted octopolar "ground-state" structures of *all* surfaces in (sc) rocksalt may be viewed as faceted (100) surfaces. For the (110) surface, this faceting has the appearance of the missing-row structure in Fig. 10(a); for the (111) surface, the partially charged - yet dipole-moment free - structure in Figs. 10(b) and 12 is obtained.

Acknowledgment. This work was supported by the U.S. Department of Energy, BES Materials Sciences, under Contract W-31-109-Eng-38.

References

[1] E. Madelung, Physik. Z. **19**, 524 (1918).

[2] P. Ewald, Ann. Physik **64**, 253 (1921).

[3] See, for example, J. H. R. Clarke, W. Smith and L. V. Woodcock, J. Chem. Phys. **84**, 2290 (1984).

[4] H. M. Evjen, Phys. Rev. **39**, 675 (1932).
[5] D. Wolf, Phys. Rev. Lett. **68**, 3315 (1992).
[6] S. W. de Leeuw, J. W. Perram and E. R. Smith, Proc. R. Soc. Lond. A 373, **27** (1980).
[7] See, for example, C. Kittel, *Introduction to Solid State Physics*, Third Edition, Wiley, 1967, p. 96.
[8] R. Lacman, Colloq. Intern. Centre Nat. Rech. Sci. (Paris) **152**, 195 (1965).
[9] N. W. Ashcroft and N. D. Mermin, *Solid State Physics*, Saunders College, Philadelphia, p. 404.
[10] D. Wolf, Solid State Ionics **75**, 3 (1995).
[11] C. R. A. Catlow, K. M. Diller and M. J. Norgett, J. Phys. C **10**, 1395 (1977).
[12] C. R. A. Catlow, I. D. Faux and M. J. Norgett, J. Phys. C **9**, 419 (1976).
[13] P. W. Tasker, Philos. Mag. A **39**, 119 (1979).
[14] A.-C. Shi and M. Wortis, Phys. Rev. B **37**, 7793 (1988).

Part II

Quantum Systems

Parallel Molecular-Dynamics Simulations of Amorphous Materials

A. Omeltchenko, A. Nakano, R.K. Kalia, and P. Vashishta

Concurrent Computing Laboratory for Materials Simulations, Department of Physics & Astronomy, and Department of Computer Science, Louisiana State University, Baton Rouge, LA 70803-4001, USA

Abstract. Using multiresolution algorithms, molecular dynamics (MD) simulations are performed on parallel machines to investigate structural correlations in porous silica. Under uniform dilation, the system develops micropores which coalesce and cause fracture as a result of percolation of the largest pore. A non-equilibrium MD approach is used to calculate thermal transport in high-temperature amorphous silicon nitride. With an increase in the density, ρ, of the system, thermal conductivity scales as $\rho^{1.5}$. Similar scaling has been observed experimentally in carbon and silica aerogels.

1. Introduction

Recent progress in designing highly efficient O(N) MD algorithms combined with the computational capabilities of powerful parallel architectures has significantly expanded the range of applications of MD simulations. It has now become possible to carry out large-scale MD simulations for real materials. In this paper, we report our recent MD simulations for porous silica and silicon nitride. These materials have extensive technological applications. Porous silica has a large thermal resistance and high optical transmission which make it an excellent alternative to CFC foamed plastic in thermal insulation of refrigerators and also a desirable material for passive solar energy collection devices.[1] Silicon nitride has numerous applications because of its strong resistance to thermal shock and corrosion, and high strength at high temperatures.[2]

2. Multiresolution Molecular-Dynamics Approach

Molecular-dynamics simulations for porous silica and silicon nitride employ empirical potentials which consist of two- and three-body terms.[3,4] To account for charge transfer effects and large polarizability of oxygen and nitrogen, the two-body potentials include long-range Coulomb and charge-dipole interactions. The two-body potentials also include steric repulsion. Three-body terms take into account covalent interactions through bond bending and stretching.

The calculation of potentials and forces uses a multiresolution approach[5] which involves a reduced cell multipole method for long-range Coulomb interaction and a multiple time-step approach to non-coulombic interactions. In addition, a separable tensorial scheme is used to efficiently compute three-body covalent interactions

Using domain decomposition, the multiresolution MD scheme has been implemented on a number of parallel platforms, such as 512-node Touchstone Delta machine at Caltech and 128-node IBM SP system at Argonne National Laboratory. On both machines, the execution time scales linearly with system size N and is inversely proportional to the number of processors. Moreover, the computation time dominates the communication time. A single MD time step for a 4.2 million atom silica system takes only 4.8 sec on the Delta machine. Figure 1 shows the performance of the multiresolution MD approach on the two parallel architectures.

Springer Proceedings in Physics, Vol. 80
Computer Simulation Studies in Condensed-Matter Physics VIII
Editors: D.P. Landau · K.K. Mon · H.-B. Schüttler

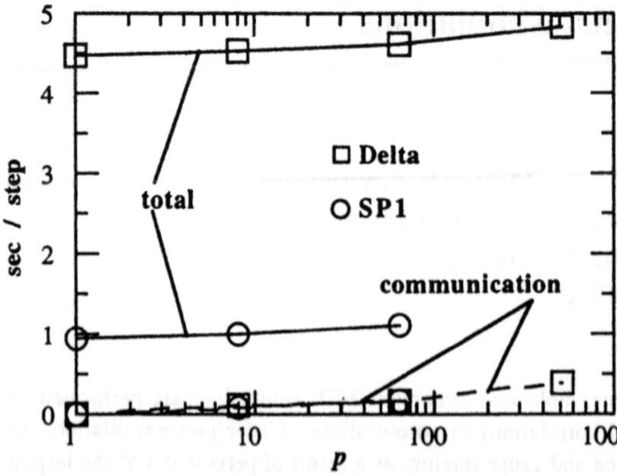

Figure 1: Execution time for the multiresolution MD program (solid curves) for 8,232p particle SiO_2 systems on p processors. Open squares and open circles are the results on the Delta and SP systems, respectively. Dashed curves represent communication overhead.

The MD simulations of porous silica were performed for densities varying from 2.2 g/cm^3 to 1.4 g/cm^3 and different system sizes up to 1.12 million atoms.[6] To prepare porous silica glasses, we melted β-cristobalite crystal by heating it to 5000K and thermalizing for several thousand time steps (the MD time step was 1 femtosecond). Then we successively cooled and thermalized systems at 4000, 3000, 2000, 1500, 600, and 300K. Porous glasses were prepared by uniformly scaling the coordinates of all the particles in the room-temperature glass by a factor of 1.02 to 1.026. After scaling, the systems were thermalized and brought to a zero-force configuration by applying the steepest-descent quench using the conjugate gradient scheme. The resulting system was heated to necessary temperatures and statistical averages were evaluated.

Amorphous Si_3N_4 systems were prepared in a similar fashion.[4] First α-Si_3N_4 crystal at a density of 3.2 g/cm^3 was melted at 5000K and then the molten system (large silicon and nitrogen diffusion was observed) was thermalized for 60,000 time steps. A 1500K glass at 3.2 g/cm^3 was produced by quenching and thermalizing the system for 45,000 time steps. Again, lower-density systems at 3.0, 2.8, 2.6, 2.4, 2.2, and 2.0 g/cm^3 were prepared by successively scaling particle coordinates and thermalizing each system for 15,000 time steps. All these systems were then cooled to 1000, 700, 500, and 300K. Each of these systems was thermalized for at least 30 ps before calculating structural and dynamical correlations.

3. Structural Transformation in Porous Silica

We observe a series of interesting morphological changes in porous silica as its density decreases.[6] Figure 2 shows snapshots of pores in a 1.12 million particle silica system at densities 1.8 and 1.4 g/cm^3. The formation of pores is first observed at a density of 1.8 g/cm^3. At lower densities, the number of pores increases and they coalesce to form larger entities. Pores are analyzed by dividing the MD box into small cubic cells of size ~ 4 Å and then we perform the cluster analysis of unoccupied cells using the breadth first search algorithm. We have investigated the morphology of pores by calculating the width, W,

$$W = \left\{ \frac{1}{N_s} \sum_{i=1}^{N_s} \left(\left| \mathbf{r}_i' - \mathbf{r}_o \right| - R \right)^2 \right\}^{1/2}, \qquad (2)$$

as a function of the pore volume, s (the number of cells within a pore times the volume of a small

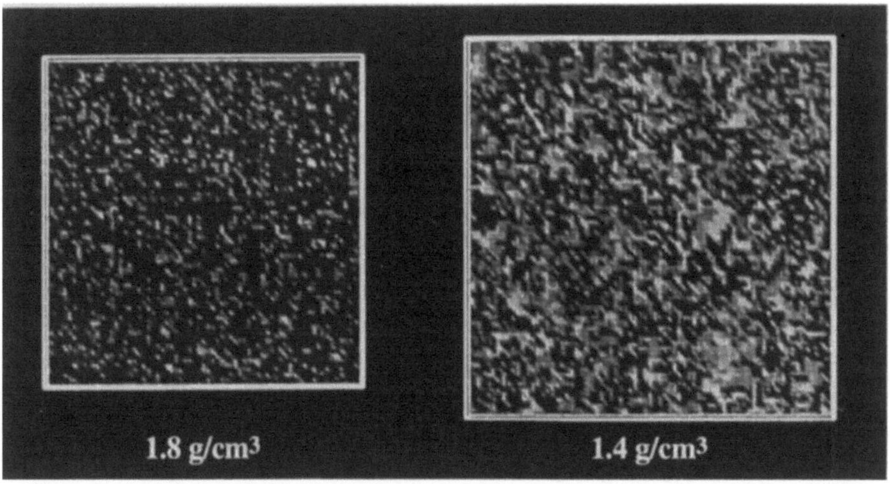

Figure 2: Snapshots of two-dimensional slices of MD configurations of silica (N = 1,119,744) at densities 1.8 g/cm^3 and 1.4 g/cm^3. Light and dark regions represent pores and silica, respectively.

cubic cell). In Eq. (2), $\{\mathbf{r}_i'\}$ denote points at the interface and $\mathbf{r}_o$ and R are the center and radius of a pore, respectively:

$$R^2 = \frac{1}{N_s}\sum_{i=1}^{N_s}\left|\mathbf{r}_i' - \mathbf{r}_o\right|^2; \quad \mathbf{r}_o = \frac{1}{N_v}\sum_{i=1}^{N_v}\mathbf{r}_i; \tag{3}$$

where $\{\mathbf{r}_i\}$ are points inside the pore.

Figure 3 shows that the radius, R, and interface width, W, of pores scale as $R \sim s^{\nu}$ and $W \sim s^{\mu}$, respectively. The exponents $\nu = 0.49 \pm 0.02$ and $\mu = 0.50 \pm 0.03$. The fractal dimension of pores, $d = 1/\nu = 2.05 \pm 0.10$. This agrees with the prediction of the percolation theory that d = 2 for lattice animals.[7] Within statistical error, we find the same values of ν and μ for porous silica at various densities between 1.7 and 1.4 g/cm^3.

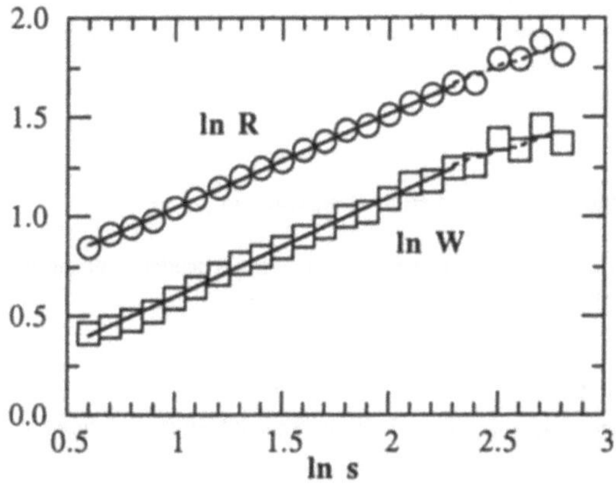

Figure 3: Variations of the average pore radius, R (open circles), and width, W (open squares), with the pore size, s, in silica at a density of 1.44 g/cm^3. Solid and dashed curves are least square fits below and above ln s = 5.3 (corresponding to R = 46 Å), respectively.

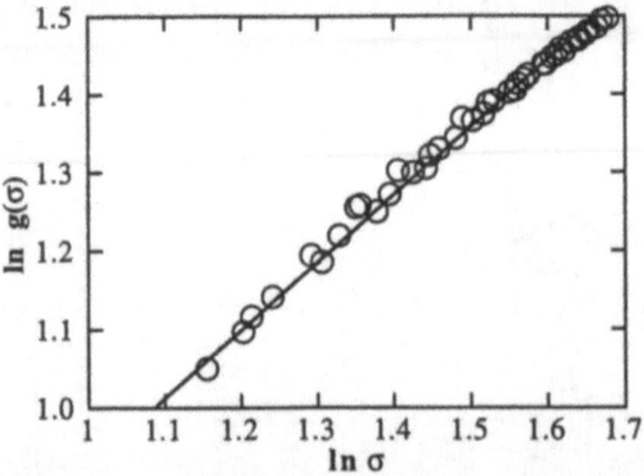

Figure 4: Height-height correlation function (open circles) versus the in-plane distance, σ, for a fracture surface resulting from pore percolation. The solid curve is the best fit, $g(\sigma) \sim \sigma^{\alpha}$, with $\alpha = 0.87 \pm 0.02$.

The average size of pores increases dramatically as the mass density approaches the critical value, ρ_c. Near ρ_c, the average pore size is given by,

$$S_{av} = \frac{\sum_s s^2 n(s)}{\sum_s s\, n(s)} \sim |\rho - \rho_c|^{-\gamma}, \tag{4}$$

where n(s) is the number of s-size pores per site. The best fit for the MD results is obtained with $\rho_c = 1.40 \pm 0.04$ and $\gamma = 1.89 \pm 0.15$, the latter being close to the universal value for percolation in three dimensions.[7]

We have also investigated the roughness of fractured surfaces resulting from percolation by calculating the height-height correlation function, $g(\sigma)$:

$$g(\sigma) = \left\langle [h(y+y_0, z+z_0) - h(y_0, z_0)]^2 \right\rangle^{1/2} \; ; \; \sigma = (y^2 + z^2)^{1/2}, \tag{5}$$

where h(y, z) is the highest vertical coordinate at the point (y, z). Figure 4 shows that the MD results for $g(\sigma)$ obey the relation, $g(\sigma) \sim \sigma^{\alpha}$ with the roughness exponent, $\alpha = 0.87 \pm 0.02$. Experimental measurements on different brittle materials obtain $\alpha \approx 0.8$ (ref. 8), as do MD simulations of two-dimensional Lennard-Jones system.[9]

4. Thermal Conductivity of Amorphous Silicon Nitride at High Temperatures

Experimental studies of amorphous solids have revealed that the temperature dependence of thermal conductivity, λ, has three distinct regimes:[10]

i) below T = 1K, λ increases as $T^{1.8-2}$;
ii) between 1 and 20K, there is a plateau in λ;
iii) beyond 30K, λ increases with temperature, approaching a limiting value at high temperatures.

The low-temperature behavior of thermal conductivity is reasonably well understood. The quadratic temperature dependence of λ below 1K is attributed to scattering of long-wavelength phonons by two-level tunneling states in glasses. However, thermal conductivity in the other two temperature regimes is far from being well-understood.

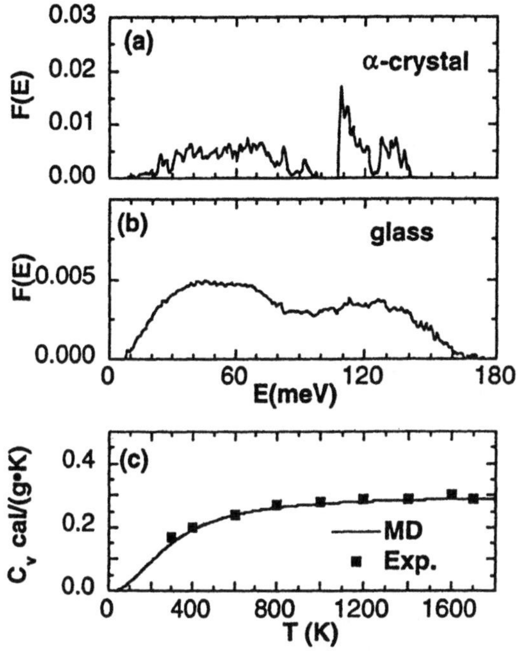

Figure 5: MD phonon density-of-states (DOS), F(E), at 3.2 g/cm^3: (a) α-crystal; and (b) glass prepared from a 4,000K melt. (c) MD and experimental specific heats for Si_3N_4 at 3.2 g/cm^3 for α-crystal.

We have investigated the nature of phonons and thermal transport in amorphous Si_3N_4. First, the system was brought to an underlying force-free configuration by applying the steepest descent quench. Then the phonon densities-of-states (DOS) was calculated by diagonalizing the dynamical matrix. In figs. 5 (a) and (b), the phonon DOS for α-crystalline silicon nitride is compared with that for a glass.[4] The calculated DOS of the α-phase has a 10 meV energy gap which is not observed in the glass. The high-energy part of the spectrum shifts from 145 meV in the α-phase to 170 meV in the glass. The specific heat of the α-crystal at normal density is calculated from the DOS, and the results along with experimental data are shown in Fig. 5 (c).[4] The MD results for the specific heat agree very well with the experimental results. Figures 6 (a) and (b) show the DOS for glasses at densities ranging from 3.2 to 2.0 g/cm^3. As the density and the connectivity of the glass is reduced, the linear part of the DOS below 15 meV grows uniformly. The maximum frequency in the DOS decreases as the density is reduced from 3.2 to 2.6 g/cm^3, and then it increases with a decrease in the density from 2.6 to 2.0 g/cm^3. The reduction in the density of the glass from 3.2 to 2.0 g/cm^3 covers two regimes. Between 3.2 and 2.6 g/cm^3 the glass stretches, but no major bond breaking occurs. When the density of the glass is reduced below 2.6 g/cm^3, the system begins to undergo decohesion and substantial number of bonds are broken. Small pores/voids develop to reduce the strain in the system. Since the nitrogen ion is large and its mass is considerably smaller than that of Si, the effect of density reduction is more pronounced in nearest neighbor N-N coordination numbers and in the average frequency of nitrogen vibrations.

Thermal conductivity of amorphous Si_3N_4 is calculated using a non-equilibrium molecular-dynamics (NEMD) approach which involves two sets of calculations starting from the same initial configuration:[11] An equilibrium MD simulation and a non-equilibrium simulation in the presence of a weak external perturbation that causes the response of the system to be proportional to the thermal

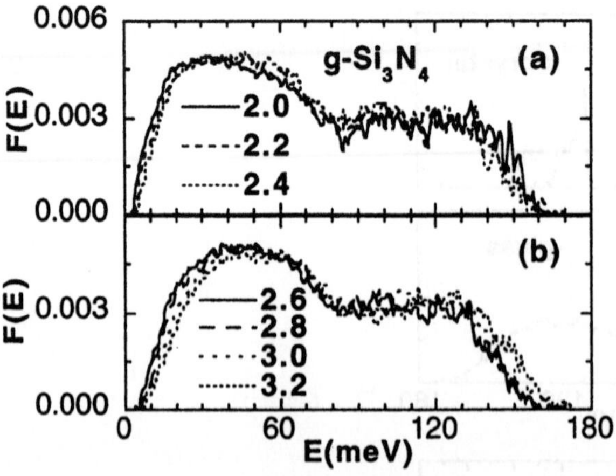

Figure 6: MD phonon DOS for glasses at densities between 3.2 and 2.0 g/cm^3

conductivity. The response to the perturbation is calculated from the difference in the heat currents of the non-equilibrium and equilibrium simulations. This subtraction procedure significantly enhances the signal-to-noise ratio of the response by reducing spurious fluctuations in the current. In the NEMD scheme we have used, the equations of motion are modified to include the coupling of the perturbation to the energy density:

$$\dot{\vec{p}}_i = \vec{f}_i + \vec{b}\,\delta\varepsilon_i - \sum_j \frac{\partial V_j}{\partial \vec{r}_i}\left(\vec{r}_{ij}\cdot\vec{b}\right) + 1/N \sum_j \sum_k \frac{\partial V_k}{\partial \vec{r}_j}\left(\vec{r}_{ij}\cdot\vec{b}\right). \tag{6}$$

where V_j is the interaction energy experienced by particle j; r_i and m_i are the position and mass of particle i, respectively; $\delta\varepsilon_i$ is the deviation of the single-particle energy from its average; and b is the strength of the perturbation. The force, F_i, is derived from, $F_i = -\nabla V$, where $V = \sum V_i$. For both equilibrium and nonequilibrium simulations, the heat current is calculated from the expression,

$$\vec{J}_q(t) = \sum_i \left[\delta\varepsilon_i \mathbf{I} - \sum_j \left(\vec{r}_i - \vec{r}_j\right)\frac{\partial V_j}{\partial \vec{r}_i}\right]\cdot\frac{\vec{p}_i}{m_i}, \tag{7}$$

where p_i is the momentum of the ith particle. The thermal conductivity in the long-wavelength limit is given by,

$$\lambda = \lim_{t\to\infty}\frac{\vec{b}\cdot\langle\vec{J}_q\rangle}{b^2} = \frac{1}{3k_B T^2\Omega}\int_0^{\infty}\left\langle\vec{J}_q(t)\vec{J}_q(0)\right\rangle dt. \tag{8}$$

The first expression on the right-hand side provides an easy and reliable evaluation of thermal conductivity which agrees with the value obtained from the equilibrium autocorrelation function method albeit with smaller statistical errors.

Figure 7 shows the temperature dependence of the thermal conductivity at a density of 2.8 g/cm^3. The thermal conductivity increases to a limiting value at high temperatures. Similar behavior is observed at other densities as well. Figure 8 shows the variation of the thermal conductivity, λ, with the density of the system at room temperature and 1200K. At these two and other temperatures as well, the MD results can be fitted to a power law, $\lambda \sim \rho^{1.5}$. It should be noted that experimental measurements[12] on carbon and silica aerogels reveal that thermal conductivity of these systems increases with the density as $\lambda \sim \rho^{1.5}$. Thus, our MD results and experimental data indicate that the density variation of thermal conductivity of porous materials is material-independent.

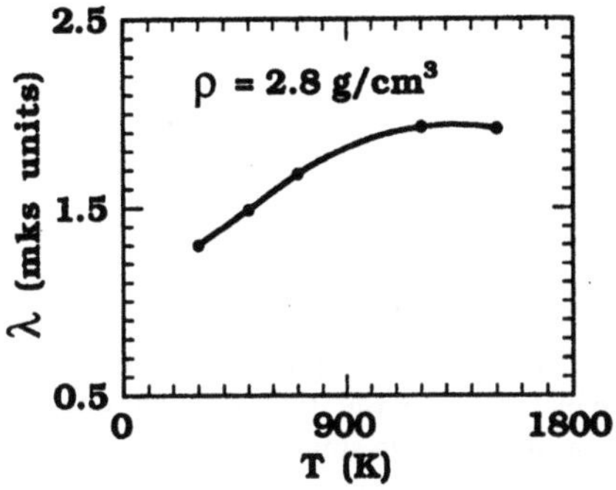

Figure 7: Temperature dependence of thermal conductivity in a-Si_3N_4.

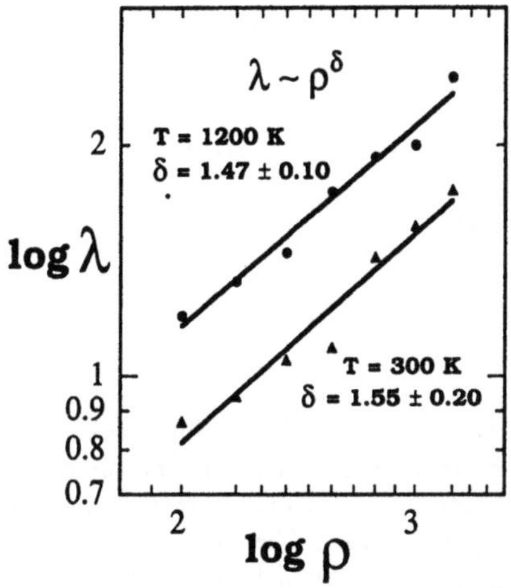

Figure 8: Density variation of thermal conductivity in a-Si_3N_4 at 300 and 1200 K.

Acknowledgments

This work was supported by DOE Grant No. DE-FG05-92ER45477, NSF Grant No. DMR-9412965, AFOSR Grant No. F49620-94-1-0444 and Louisiana LEQSF. Computations were performed in the Concurrent Computing Laboratory for Materials Simulations at Louisiana State University using the facilities acquired with LEQSF Equipment Enhancement Grants, and also on the IBM SP system at Argonne National Laboratory and the Touchstone Delta machine at Caltech.

References

1. J. Fricke, J. Non-Cryst. Solids 121, 188 (1990).
2. Silicon Nitride - 1, eds. S. Somiya, M. Mitomo, and M. Yoshimura, (Elsevier Science, NY, 1990); Silicon Nitride by P. T. B. Shaffer and A. Goel, (Advanced Refractory Technologies Inc, Buffalo, NY, 1993).
3. P. Vashishta, R. K. Kalia, J. P. Rino, and I Ebbsjo, Phys. Rev. B 41, 12197 (1990).
4. P. Vashishta, R. K. Kalia, and I. Ebbsjö, to be published.
5. A. Nakano, R. K. Kalia, and P. Vashishta, Comput. Phys. Commun. 83, 197(1994).
6. A. Nakano, R. K. Kalia, and P. Vashishta, Phys. Rev. Lett. 73, 2336 (1994).
7. D. Stauffer, Introduction to Percolation Theory, (Taylor and Francis, London).
8. K. J. Maloy et al., Phys. Rev. Lett. 68, 213 (1992).
9. F. F. Abraham et al., Phys. Rev. Lett. 73, 272 (1994).
10. P. B. Allen and J. L. Feldman, Phys. Rev. Lett. 62, 645 (1989).
11. D. J. Evans, Phys. Lett. 91A, 457 (1982).
12. X. Lu et al., Science 255, 971 (1991).

Monte-Carlo Calculation of Quantum $J_1 - J_2$ Model on the Square Lattice

S. Miyazawa and S. Homma

Faculty of Engineering, Gunma University, Kiryu 376, Japan

Abstract: We used the modified Decoupled Cell method (mDCM) of quantum Monte Carlo simulation to calculate the thermodynamic properties and spin configurations of the frustrated $J_1 - J_2$ model on the square lattice by increasing the frustration parameter $\alpha = \frac{J_2}{J_1}$ from 0 to 1. The size N of a system used in Monte Carlo simulation in this study is 32×32. We found that for small values of α the Neel state is a ground state spin configutation of this model system, whereas for $\alpha > 0.6$ a collinear state is its ground state configuration instead of the Neel state.

1. Introduction

Since the discovery of high T_c-superconductor the two-dimensional quantum spin ($s = \frac{1}{2}$) system have received much attention experimentally as well as theoretically [1]. Particularly the frustrated $J_1 - J_2$ model on the square lattice (hereafter we call it $J_1 - J_2$ model) have attracted much interest because of the suggestion that the effect of the hole doping in the high T_c superconductor may be simulated by the introduction of frustrations into the antiferromagnetic Heisenberg model with the nearest neighbor interaction [2]. In addition it is expected that this model exhibits the interplay between the frustration and the quantum fluctuation, an important problem in the low-dimensional quantum spin systems. The main problems there are which is the ground state spin configuration, and how are the thermodynamic propeties such as the specific heat, magnetic susceptibility and the spin pair correlation functions when the frustration parameter α increases from 0 to 1.

In the case of the classical $J_1 - J_2$ model the ground state is the Neel-ordered state for $\alpha < 0.5$, and is the continuously degenerate four-sublattice state for $\alpha > 0.5$. At $\alpha = 0.5$, the two classical states are degenerate. In the quantum case the Neel order is believed to be the stable ground state for $\alpha \ll 0.5$. It has been proved that the nonexistence of the twisted ordered state [3]. For the case where α is near 1, it is supposed that a collinear state is the ground state spin configuration [4]. In the region, in which α is near 0.5 the ground state configuration is still remained to be solved. For the theoretical studies the spin wave theory, ordinary as well as modified [5-6], Schwinger-boson mean-field theory [7], finite lattice study [8-13] and series expansions [14-15] are to be noted. However their results for the ground state spin configuration are different to each other. For example some works predicted a disordered or spin liquid state at around $\alpha = 0.5$, and the others claimed that the classical Neel state is a ground state configuration even in the quantum mechanical $J_1 - J_2$ model. Numerical studies of finite lattices based on the exact diagonalization method gave the results that at 0 K the Neel state gradually decreases with an increase of α and at about $\alpha = 0.6$ this state disappear [16]. However the number of lattice sites included in those studies were at most 36 and is not enough to

Springer Proceedings in Physics, Vol. 80
Computer Simulation Studies in Condensed-Matter Physics VIII
Editors: D.P. Landau · K.K. Mon · H.-B. Schüttler

extract definite conclusions of the physical quantities of the system in the thermodynamic limit.

To overcome this difficulty and to extend the study at finite temperatures, a quantum Monte Carlo calculation was used by invoking the generalized Trotter formula (Suzuki-Trotter formula), and gave informations on the magnetic properties of this model [16]. However there exists a serious negative sign problem there and also lattice sizes used there were small one. Therefore it is desirable to invoke a quantum Monte Carlo method which is free from the negative sign problem and also is able to extend a lattice size with ease. The Decoupled Cell method (DCM) and the modified Decoupled Cell method (mDCM), which were proposed in 1984 and 1993, and used extensively since then, is the one which is free from these difficulties [17-19].

It is the purpose of the paper to study the thermodynamic properties of the quantum $J_1 - J_2$ model on the square lattice by applying mDCM and show the results obtained. In the next section we give a detailed account of DCM and mDCM. In the section 3 we show the results obtained by applying mDCM to $J_1 - J_2$ model. The last section is devoted to discussion. The preliminary report of this work was published in ref.[23].

2. The Decoupled and the modified Decoupled Cell method

To explain the basic idea of DCM let us consider a quantum spin system ($s = 1/2$) whose Hamiltonian is given by H. The state of the i-th site can be specified by a variable $s = \pm 1/2$, and the state of the total system by an N-dimensional state vector $|S> = |s_1, s_2, ..., s_N>$ whose i-site state is s_i. The probability of the state S in the canonical distribution is given by

$$P(S) = < S|\exp(-\beta H)|S > /Z, \tag{1}$$

where $\beta = (1/kT)$ and Z stands for the partition function of a system. As in the classical Metroplis method, the condition of irreducibility and recurrency of the Markov chain is satisfied by assigning positive transition probabilities between states that are different from each other only at one site $i (i = 1, 2, ..., N)$. The problem in the quantum mechanical case is how to obtain adequate transition probabilities consistent with the condition of the detailed balance at equilibrium

$$P(S)W(S \to S') = P(S')W(S' \to S) \tag{2}$$

where W$(S \to S')$ is the transition probability from a spin configuration S to S'.

Let $L_i(\nu)$ be a set of sites whose distance from the i-th site does not exceed a certain integer ν and $\overline{L}_i$ be a set of all sites not belonging to $L_i(\nu)$. We call such $L_i(\nu)$ the decoupled cell (DC) of radius ν with its center at the i-site. Let S_i denote the state of $L_i(\nu)$ excepting the i-site and $\overline{S}_i$ denote the state $\overline{L}_i$. The state of the total system can then be written as $S = (s_i, S_i, \overline{S}_i)$. The transition probability between $S = (s_i, S_i, \overline{S}_i)$ and $S' = (-s_i, S_i, \overline{S}_i)$ can be obtained from Eq.(1) if one knows the value of

$$q(S) = \frac{P(S)}{P(S')} = \frac{< S|\exp(-\beta H)|S >}{< S'|\exp(-\beta H)|S' >}: \tag{3}$$

Let $H(\nu, i)$ be the Hamiltonian of a DC which is obtained from H by deleting all the terms containing operators of $\overline{L}_i$. The basic ingredient of DCM is to approximate Eq.(3) by

$$q^{(\nu)}(S_i) = \frac{< s_i, S_i | \exp(-\beta H(\nu, i)) | s_i, S_i >}{< -s_i, S_i | \exp(-\beta H(\nu, i)) | -s_i, S_i >}. \tag{4}$$

This approximation is based on the presumption:

The dependence of q on $\overline{S}_i$ gradually decreases by increasing ν.

The right hand side of Eq.(4) can be obtained readily by solving the eigenvalue problem of $H(\nu, i)$ by computer. Let ($E_n, \phi_n; n = 1, 2, .., f$ = a number of spin states of DC) be eigenvalues and eigenfunctions of $H(\nu, i)$. With their use Eq.(4), $q^{(\nu)}(S_i)$ can be rewritten as

$$q^{(\nu)}(S_i) = \frac{\sum_n | < s_i, S_i | \phi_n > |^2 \exp(-\beta E_n)}{\sum_n | < -s_i, S_i | \phi_n > |^2 \exp(-\beta E_n)}. \tag{5}$$

The transition probability W_{DC} defined in DCM is given by

$$W_{DC}(-s_i \to s_i) = max[1, q^{(\nu)}(S_i)]. \tag{6}$$

The physical meaning of replacing Eq.(3) by Eq.(4) was fully discussed by Matsuda et al. [20]. If we use Eq.(6) as the transition probability from the state $S' = (-s_i, S_i, \overline{S}_i)$ to the state $S = (s_i, S_i, \overline{S}_i)$ in Monte Carlo calculation, we are able to obtain the Markov chain of the given quantum mechanical system using the Metroplis algorithm. Thus DCM gives a natural extension of the classical Monte Carlo method to quantum systems. The detailed computational procedure of DCM is given in references [17,21] . We applied DCM to one-dimensional XY model and compared the results with the exact one [22] which is derived analytically and found that by increasing the size of DC the obtained results gradually approach the exact one except for very low temperature region, where the calculated value of an internal energy gives a negative specific heat and the perpendicular susceptibility increases abruptly contrary to the fact that the exact analytical solution is finite at absolute zero of temperature [17,21].

These difficulties in the low temperature region might be attributed to the breakdown of the detailed balance originating from the finiteness of a DC used there. In order to remove this difficulty there are two possibilities. The first is to enlarge the size of DC, but it is limited by the capability of computer power. The second is to improve DCM to recover the detailed balance, extending the basic concept of DCM for a certain size of DC. As we pointed out previously, DCM could be regarded as a natural extension of the classical Monte Carlo method to quantum systems. However, it should be noted that the transition probability defined by Eq.(6) does not satisfy the detailed balance. The probability $W_{DC}(s_i \to -s_i)$ is a function of the neighbouring spins in the cell, which includes a finite number of spins. However, when we calculate the transition probability of some other spin in the same cell, we introduce W_{DC} in the form of Eq.(6) independently. If the cell Hamiltonians $H(\nu, i)$ commute with each other, the detailed balance is satisfied automatically as far as it is satisfied locally. Thus Eq.(6) gives correct transition probabilities for classical systems. But in quantum systems where $H(\nu, i)$ do not commute with each other, the transition probability defined at each lattice site independently does not satisfy the detailed balance. The flip of a spin s_k causes a change of all the transition probabilities for which $H(\nu, i)$ includes s_k. Thus the changes cannot be reduced to that of $W_{DC}(s_i \to -s_i)$. From this point of view, in determining the transition probability it is important to include not only the DC whose central site is i, but all the DC which include the i-site. Here we reformulate DCM taking into account the above considerations.

First we decompose a system into identical cells (decoupled cell; DC), whose shape and size are given. The way of a decomposition is not unique. It depends on the size and shape of DC. With a decomposition, labeled by j and a k-th cell, we associate the cell Hamiltonian $H_n(j,k)$, where n is the number of lattice sites included in each cell (DC). Then the Hamiltonian of a system is written as a sum of $H_n(j,k)$ as

$$H = \frac{1}{r(n)} \sum_j (\sum_k H_n(j,k)), \tag{7}$$

where the sum over k means the sum over all cells on a lattice and that over j means the sum over all different decompositions. The symbol $r(n)$ denotes the number of different decompositions.

Using Eq.(7) the probability of a certain spin configuration $|S, s_i>$ is given by

$$P(S, s_i) = \frac{1}{Z} < S, s_i | \exp(-(\frac{\beta}{r(n)} \sum_j (\sum_k H_n(j,k))) | S, s_i >, \tag{8}$$

where Z is the partition function of a system. The ket (bra)$|S, s_i>$ ($<S, s_i|$) represents a spin configuration of a system, in which the i-th spin is s_i, with $s_i = \pm(1/2)$. We approximate Eq.(8), invoking the basic concept discussed above, as

$$P(S, s_i) = \frac{1}{Z} \Pi_j \Pi_k < \sigma(j,k) | \exp(-(\frac{\beta}{r(n)} H_n(j,k))) | \sigma(j,k) >, \tag{9}$$

where $|\sigma(j,k)>$ represents a spin state of a cell (DC) labeled by (j,k).

The transition probability $W(s_i \rightarrow -s_i)$ is defined by

$$W(s_i \rightarrow -s_i) = max[1, \frac{P(S, -s_i)}{P(S, s_i)}], \tag{10}$$

where $P(S, \pm s_i)$ is defined by Eq.(8). We approximate Eq.(8) , by substituting Eq.(9) into $P(S, \pm s_i)$ in Eq.(10) , to obtain W_{DC} in the modified Decoupled Cell Method (mDCM) as

$$\begin{aligned} & W_{DC}(s_i \rightarrow -s_i) \\ = \; & max[1, \frac{\Pi_j \Pi_k < \sigma(j,k), -s_i | \exp(-\frac{\beta}{r(n)} H_n(j,k)) | \sigma(j,k), -s_i >}{\Pi_j \Pi_k < \sigma(j,k), s_i | \exp(-\frac{\beta}{r(n)} H_n(j,k)) | \sigma(j,k), s_i >}]. \end{aligned} \tag{11}$$

The product over k in Eq.(11) must be over all cells which include the i-site; the total number of such cells is equal to the number of spins in a cell, that is n. Here.it must be noted that in DCM only the cell whose center is i-site is taken into account in Eq.(11) . If the Hamiltonian H of a system consists of only a nearest neighbour coupling, Eqs.(10) and (11) coincide with those of classical ones in the classical limit. However it should be noted that in one-dimensional lattice when the Hamiltonian H includes the second-neighbour interaction beside the nearest neighbour interaction, each second-neighbour interaction is included in Eq.(9) by a factor $\frac{n-2}{n-1}$, whereas the nearest neighbour interaction is included in Eq.(9) by $\frac{n-1}{n-1} = 1$. Thus Eq.(11) does not give a correct expression in the classical limit in this case (in one-dimensional lattice with a second-neighbour interaction). In two-dimensional lattice proper size and shape of a cell (DC) depend on the type of lattice and the range of interactions. For the detailed account of the possible decompositions of a certain two-dimensional lattice with nearest- and second-neighbour interactions, into DC of the given shape and size , the reader is referred to ref.[18]. Thus we complete the modification of DCM and have a modified Decoupled Cell method (mDCM) .

3. $J_1 - J_2$ model

The Hamiltonian of the quantum $J_1 - J_2$ model is defined as

$$H = 2J_1 \sum_{i,j}^{nn} \mathbf{S}_i \cdot \mathbf{S}_j + 2J_2 \sum_{i,j}^{nnn} \mathbf{S}_i \cdot \mathbf{S}_j. \tag{12}$$

Here J_1 and J_2 are the nearest-neighbor and next-nearest-neighbor interactions, both of which are antiferromagnetic ($J_1, J_2 > 0$) on a square lattice. $\mathbf{S}_i$ is a quantum spin operator ($S = 1/2$) of the i-th lattice site. The Hamiltonian is isotropic, so that we take the z-axis as the axis of a diagonal representation. The thermodynamic quantities we calculate here are an internal energy, specific heat, total magnetization, susceptibility of a total system and also spin configurations at various temperatures for the given frustration parameter α. We did not calculate staggered magnetizations and susceptibilities, for at present we do not know the spin configuration of the Hamiltonian for a given α. The size of the decoupled cell (DC) used in here is depicted in Fig.1, where the interaction on each edge is $\frac{J_1}{2}$ instead of J_1. This is because coupling constants on the edge of DC are split equally into two neighboring DC in the cell decomposition of the Hamiltonian. The number n of lattice sites included in DC in Fig.1 is 9. The total number N of lattice sites in the system used in the present Monte Carlo calculations are 32×32 with a periodic boundary condition.

Calculations were performed using the Metropolis Monte Carlo procedure. The run was taken at $k_BT = 2.0J_1$, starting from a random configuration. The first 1000 Monte Carlo steps (MCS) were used to make the system in thermal equilibrium and the following 10000 MCS were used to calculate the thermodynamic quantities and the spin configurations. The system is then cooled in steps down to $k_BT = 0.1J_1$. At each temperature the initial configuration was taken from that of the final one of the previous temperature and first 1000 MCS were used to get the thermal equilibrium.

In calculating the energy of the system we have used the following expression

$$\epsilon = \frac{1}{Nr(n)} \sum_{j=1}^{r(n)} \left(\sum_k < H_n(j,k) >\right), \tag{13}$$

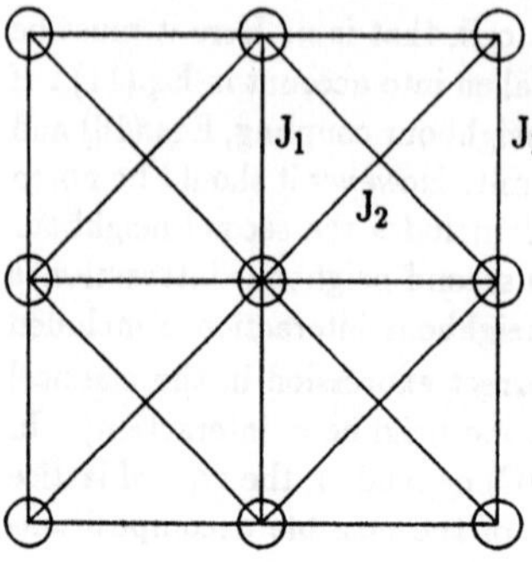

Fig.1 A size of a decoupled cell (DC).

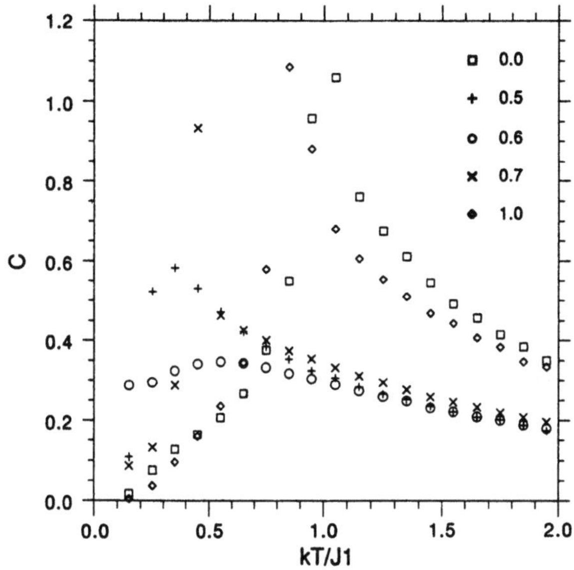

Fig.2 Specific heat of the quantum $J_1 - J_2$ model ($J_1 - J_2$ model), where numbers in the figure represent the numerical value of α.

where $< >$ denotes the average with respect to the cell Hamiltonian $H_n(j,k)$ and $r(n)$ means the number of different decompositions. The sum over k and j means the sum of cells over a whole lattice and the sum over all the different decompositions. For the DC depicted in Fig.1 $r(n=9)$ is 4 . The specific heat c is calculated as

$$c = \frac{\Delta\epsilon}{\Delta T} \tag{14}$$

where $\Delta\epsilon$ and ΔT are energy and temperature differences, respectively. The total magnetization M_z along z-axis is

$$M_z = < \sum_i S_i^z > \tag{15}$$

Here $< \Omega >$ denotes the canonical average of Ω . The mean square magnetization of a system, which is written as $m_z(2)$, is defined as

$$m_z(2) = \frac{1}{N} < (\sum_i S_i^z)^2 > \tag{16}$$

Thus, the magnetic susceptibility χ of a system is calculated as

$$\chi = \frac{1}{Nk_BT} < (\sum_i S_i^z)^2 > = \frac{m_z(2)}{k_BT}. \tag{17}$$

The spin pair correlation function $C^z(r)$, which is defined as

$$C^z(r) = < S_i^z S_{i+r}^z > \tag{18}$$

was calculated along x- and y-axis.

In Fig.2 we show the result of the specific heat for various values of α. There are two things to be noticed. The peak in the specific heat is sharp in the extreme cases

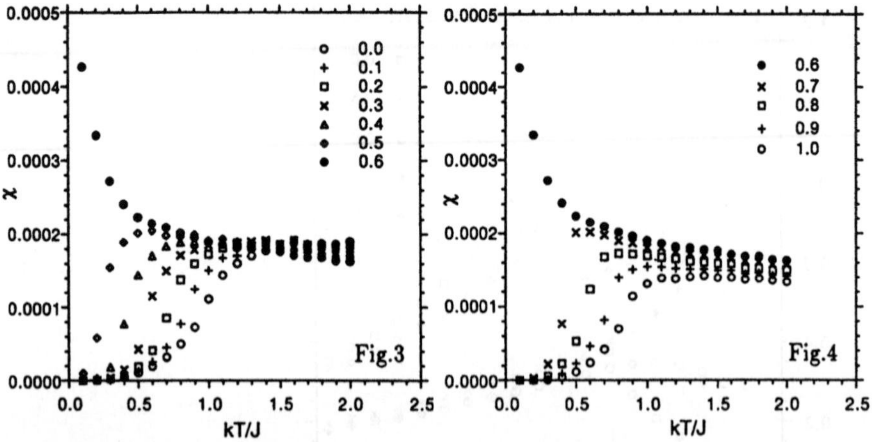

Fig.3 The magnetic susceptibility χ of $J_1 - J_2$ model for $\alpha < 0.6$, where numbers in the figure represents α.

Fig.4 The magnetic susceptibility χ of $J_1 - J_2$ model for $\alpha > 0.6$, where numbers in the figure represents α.

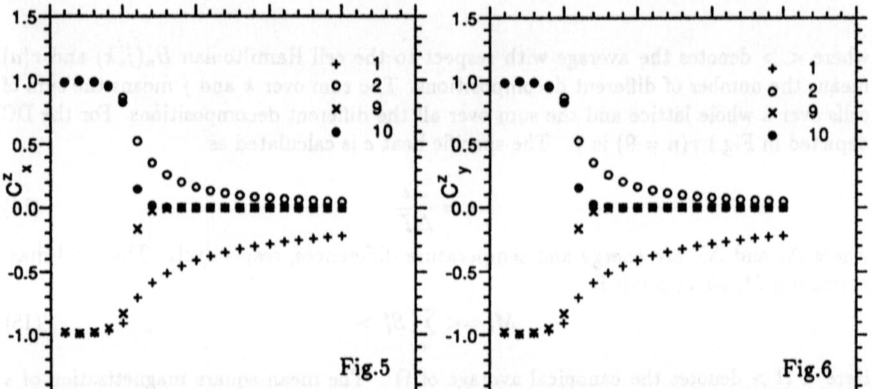

Fig.5 The spin correlation function $C^z(r)$ along x-axis, where numbers in the figure mean distance in unit of lattice constant. ($\alpha = 0.3$)

Fig.6 The spin correlation function $C^z(r)$ along y-axis , where numbers in the figure mean distance in unit of lattice constant. ($\alpha = 0.3$)

of $\alpha = 0$ and 1, and appears in the high temperature region. This sharpening of the peaks might be interpreted as the gradual ordering of the system into the Neel state and the collinear state, respectively. Second, the height and the position of peaks in the intermediate values of α become low and move to the low temperature region. Thus, the

Fig.7 The spin configuration of $J_1 - J_2$ model for $\alpha = 0.3$ at $k_B T = 0.1 J_1$.

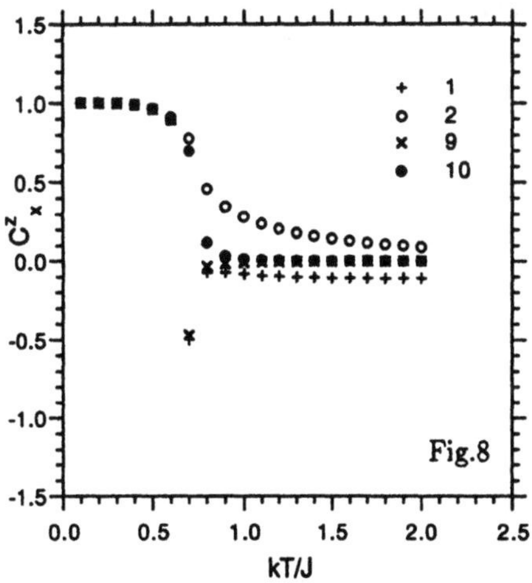

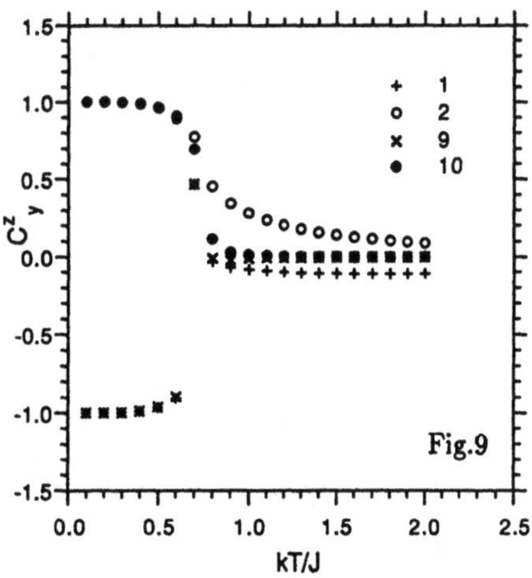

Fig.8 The spin correlation function $C^z(r)$ along x-axis, where numbers in the figure mean distance in unit of lattice constant. ($\alpha = 0.9$)

Fig.9 The spin correlation function $C^z(r)$ along y-axis , where numbers in the figure mean distance in unit of lattice constant. ($\alpha = 0.9$)

peak there indicates that one can not expect any finite temperature transition. For the total magnetization M_z we always have observed zero as it should be. In Figs.3 and 4 we show the result of the susceptibility χ. From this figure we see that in the case of $\alpha = 0.6$, χ increases in the low temperature region , whereas above and below $\alpha = 0.6$, χ decreases to zero, showing that there exists ordered states of a spin configuration for those values of α. In order to obtain the spin configurations we have calculated the spin pair correlation functions defined by Eq.(18) along the x- and y-axis. In Figs.5 and 6 we show $C^z(r)$ for $\alpha = 0.3$ along x- and y-axis. The numbers in each figure indicates r. From them we obtain the spin configuration, which is depicted in Fig.7 at $\frac{kT}{J_1} = 0.1$. Thus we have Neel State for $\alpha = 0.3$. We have observed similar behaviour for $\alpha = 0.0$, 0.1,0.2, 0.4 and 0.5. In Figs.8 and 9 we show $C^z(r)$ for $\alpha = 0.9$. From them we have a collinear spin

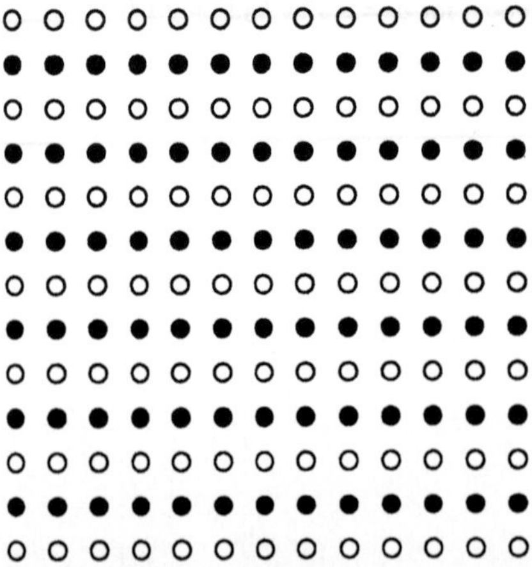

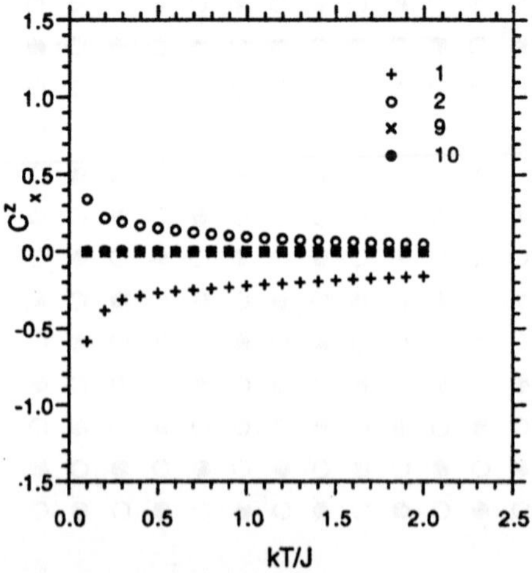

Fig.10 The spin configuration of $J_1 - J_2$ model for $\alpha = 0.9$ at $k_B T = 0.1 J_1$.

Fig.11 The spin correlation function $C^z(r)$ along x-axis, where numbers in the figure mean distance in unit of lattice constant. ($\alpha = 0.6$)

Fig.12 The spin configuration of $J_1 - J_2$ model for $\alpha = 0.6$ at $k_B T = 0.1 J_1$.

configuration for this value of α. This is shown in Fig.10. For $\alpha > 0.6$ we have obtained a collinear spin configuration. From these observations we conclude that the ground state spin configuration of $J_1 - J_2$ model is Neel state for $\alpha < 0.6$ and a collinear state for $\alpha > 0.6$. At α= 0.6 we show in Fig.11$C^z(r)$ along x-axis. ($C^z(r)$ along y-axis is the same as that of x-axis.)

In Fig.12 we show the spin configuration at $\frac{kT}{J_1} = 0.1$, from which we see that there are seeds for the growth of Neel and collineat spin configuration, respectively.

4. Discussion

In this study we have applied mDCM, which does not have difficulties of negative sign problem, to the quantum $J_1 - J_2$ model on a square lattice and calculated the thermodynamic quantities such as an internal energy, specific heat and a magnetic susceptibility as well as spin configurations at various values of α and temperature T.

We have obtained the results that for small values of α a spin configuration is the Neel ordered state and at $\alpha = 0.5$, at which a spin configuration is degenerate in the classical case, it is still the Neel state. For $\alpha > 0.6$ we have observed that the spin configuration in the low temperature region is a collinear. At present it is not certain that at $0K$ there exits a phase transition from the Neel state to a collinear state by increasing α, or not.. In order to obtain the definite conclusion on the behavior of the system near $\alpha = 0.6$ in the low temperature region, we have to enlarge the size of DC as well as a size of a system. This will be the next problem.

References

[1] For a review, see E.Manousakis, Rev. Mod. Phys. **63** (1991) 1

[2] M.Inui, S.Doniach and M.Gabay, Phys. Rev. **B38** (1988) 6631

[3] Y.Saika and H.Nishimori, J. Phys. Soc. Jpn. **61** (1992) 3086

[4] K.Kubo and T.Kishi, J. Phys. Soc. Jpn. **60** (1990) 567

[5] J.Igarashi, J. Phys. Soc. Jpn. **62** (1993) 4449

[6] T.Aoki, J. Phys. Soc. Jpn. **63** (1994) 2356 and the references cited therein.

[7] F.Mila, D.Poilblanc and C.Bruder, Phys. Rev. **B43** (1991) 7891

[8] E.Dagotto and A.Moreo, Phys. Rev. Lett. **63** (1989) 2148

[9] K.Sano, I.Doi and K.Takano, J. Phys. Soc. Jpn. **60** (1991) 3807

[10] H.J.Schulz and T.A.L.Ziman, Europhys. Lett. **18** (1992) 355

[11] J.Richter, Phys. Rev. **B47** (1993) 5794

[12] Q.F.Zhong and S.Sorella, Europhys. Lett. **21** (1993) 629

[13] N.B.Ivanov and J.Richter, J. Phys.: Condens.Matter **6** (1994) 3785

[14] M.P.Gelfand, R.R.P.Singh and D.A.Huse, Phys. Rev. **B40** (1989) 10801

[15] I.G.Gochev, Phys. Rev. **B49** (1994) 9594

[16] T.Nakamura and N.Hatano, J. Phys. Soc. Jpn. **62** (1993) 3062

[17] S.Homma, H.Matsuda and N.Ogita, Prog. Theor. Phys. **72** (1984) 1245
75 (1986) 1058

[18] S.Miyazawa, S.Miyashita, M.S.Makivic and S.Homma,
Prog. Theor. Phys. **89** (1993) 1167

[19] S.Homma, in *Quantum Monte Carlo Methods in Condensed Matter Physics* , ed. M.Suzuki (World Scientific, Singapore, 1993) p.163

[20] H.Matsuda, K.Ishii, S.Homma and N.Ogita, Prog. Theor. Phys. **80** (1988) 583

[21] S.Homma, K.Sano, H.Matsuda and N.Ogita, Prog. Theor. Phys. **S87** (1986) 127

[22] S.Katsura, Phys. Rev. **57** (1962) 1508

[23] S.Miyazawa and S.Homma, Phys. Letters **A193** (1994) 370

Evaluation of Dynamical Spectra for $T = 0$ Quantum Monte-Carlo Simulations: Hubbard Lattices and Continuous Systems

J.J. Deisz[1], *W. von der Linden*[2], *R. Preuss*[3], *and W. Hanke*[3]

[1]Department of Physics, Georgetown University, Washington, DC 20057, USA
[2]Max-Planck-Institut für Plasmaphysik, EURATOM Association, D-85740 Garching b. München, Germany
[3]Institut für Theoretische Physik, Universität Würzburg, Am Hubland, D-97074 Würzburg, Germany

Abstract. Zero-temperature dynamical spectra for Hubbard lattices and simple atoms are obtained using ground state projection quantum Monte Carlo and the maximum entropy method. Once zero-temperature imaginary-time correlations are identified and calculated, maximum entropy deconvolutions produce results of similar quality as those obtained from finite temperature quantum Monte Carlo for which this method has been frequently employed.

1. Dynamical spectra from quantum Monte Carlo and the maximum entropy method.

Quantum Monte Carlo (QMC) algorithms have been developed for a variety of systems and models, including the electron gas model, atoms and molecules, the Heisenberg model, the Anderson impurity model, and Hubbard lattice models. To within statistical precision, QMC calculations directly provide total energies, spatial correlations, and static ($\omega = 0$) susceptibilities. Dynamical quantities ($\omega \neq 0$) are not calculated directly, though they are necessary for interpreting electrical resistivity measurements, neutron scattering, photoemission, and other dynamical probes. Recently, an algorithm for obtaining dynamical properties from finite-temperature quantum simulations was developed [1, 2]. It is shown below, via calculations for a one-dimensional Hubbard lattice model and hydrogen and helium atoms, that this algorithm is applicable to zero-temperature QMC simulations.

The finite-temperature algorithm begins with QMC evaluation of imaginary-time operator expectation values, like the single-particle Green's function,

$$G(\mathbf{k},\tau) = - < T_\tau e^{(\hat{H}-\mu\hat{N})\tau}\, \hat{c}_{\mathbf{k},\sigma}\, e^{-(\hat{H}-\mu\hat{N})\tau} \hat{c}^\dagger_{\mathbf{k},\sigma} > . \tag{1}$$

The Feynman path integral formulation of finite temperature QMC algorithms [3] expresses $G(\mathbf{k},\tau)$ as a configuration-weighted sum. Importance sampling of configurations produces numerical results for $G(\mathbf{k},\tau)$ on a discrete imaginary-time grid $\tau = 0, \Delta\tau, \cdots, \beta$.

The dynamical content of $G(\mathbf{k},\tau)$ is described by the single-particle spectral function, $A(\mathbf{k},\varepsilon)$,

Springer Proceedings in Physics, Vol. 80
Computer Simulation Studies in Condensed-Matter Physics VIII
Editors: D.P. Landau · K.K. Mon · H.-B. Schüttler

$$A(\mathbf{k},\varepsilon) = \frac{1}{\mathcal{Z}}\sum_{s,s'} e^{-\beta(E_s-\mu N_s)}(1+e^{-\beta\varepsilon})\,|<s|\hat{c}_{\mathbf{k},\sigma}|s'>|^2 \times\delta(\varepsilon-(E_{s'}-E_s-\mu)), \tag{2}$$

through the integral transformation

$$G(\mathbf{k},\tau) = -\int_{-\infty}^{\infty}\frac{e^{-\varepsilon\tau}}{1+e^{-\varepsilon\beta}}\,A(\mathbf{k},\varepsilon)\,d\varepsilon. \tag{3}$$

However, deconvolution of Eq. (3) to obtain $A(\mathbf{k},\varepsilon)$ is mathematically ill-posed since QMC results for $G(\mathbf{k},\tau)$ are available on a finite grid of τ-points and contain statistical errors.

The maximum entropy method (MEM) is the *only* Bayesian probability theory based approach for systematically obtaining approximate solutions for inferential problems where there is uncertainty in the data [4, 5]; it has been applied successfully to invert Eq. (3) [1, 2]. MEM yields the most probable $A(\mathbf{k},\varepsilon)$ in light of the QMC data and additional prior knowledge, such as positivity and sum-rules [6]. The central quantity of MEM is the relative entropy,

$$S(A,m) = -\int_{-\infty}^{\infty} d\varepsilon\,[A(\mathbf{k},\varepsilon)\ln(A(\mathbf{k},\varepsilon)/m(\varepsilon)) - A(\mathbf{k},\varepsilon)+m(\varepsilon)], \tag{4}$$

which is the log-probability derived from a Poisson-process relative to the default model, $m(\varepsilon)$, which is chosen on the basis of perturbation theory [7], sum-rules [6, 8], or simply taken as constant. If data is absent or too poor to be informative, MEM yields $A(\mathbf{k},\varepsilon) = m(\varepsilon)$. The balance between entropic 'prior' probability and data constraints is determined self-consistently [4, 5].

MEM deconvolutions have become a standard tool for analysis of data from finite-temperature QMC simulations [9]. Systems studied using MEM include Anderson impurity models [7], Hubbard lattices [6, 10], and Heisenberg antiferromagnets [11, 12]. However, the usefulness of the maximum entropy method for obtaining dynamics from zero-temperature simulations has largely remained unexplored.

Prescriptions have been developed for obtaining imaginary-time correlation functions from zero-temperature simulations for Hubbard lattices [13] and two-electron atoms and molecules [14]. To see that the dynamical content of these correlation functions is similar to its finite temperature counterpart, consider an operator $\hat{Q}$.

$$S_{QQ}(\tau) \equiv <\hat{Q}^\dagger(\tau)\,\hat{Q}(\tau=0)> \equiv <\psi_o|e^{\tau\hat{H}}\,\hat{Q}^\dagger\,e^{-\tau\hat{H}}\,\hat{Q}|\psi_o> \tag{5}$$

has the spectral decomposition

$$S_{QQ}(\tau) = \sum_m e^{-(E_m-E_o)\tau}\,|<\psi_m|\hat{Q}|\psi_o>|^2. \tag{6}$$

It is simple to show that the dynamical structure factor or spectral function,

$$S_{QQ}(\omega) = \sum_m | < \psi_m|\hat{Q}|\psi_o > |^2 \delta(\omega - (E_m - E_o)), \tag{7}$$

is related to $S_{QQ}(\tau)$ by a Laplace transform,

$$S_{QQ}(\tau) = \int_0^\infty e^{-\omega\tau} S_{QQ}(\omega)\, d\omega. \tag{8}$$

Thus, once imaginary-time correlation functions are evaluated, the extraction of dynamical information via Eq. (8) is essentially equivalent to obtaining dynamical information at finite temperature. In Section 2 it is demonstrated that the zero-temperature dynamical method suggested for Hubbard lattices [13] is as effective as the finite-temperature calculations. In Section 3 dynamical calculations for continuous systems are placed in the QMC+MEM framework; Results for atomic hydrogen and helium show the intrinsic energy resolution of the QMC data.

2. $T = 0$ dynamics for Hubbard lattice models.

$\exp(-\beta\hat{H})$ projects ψ_o from a trial state ψ_T when β is large:

$$e^{-\beta\hat{H}}|\psi_T > \; = \; \sum_m e^{-\beta E_m} < \psi_m|\psi_T > |\psi_m > \tag{9}$$

$$\simeq \; e^{-\beta E_o} < \psi_o|\psi_T > |\psi_o > . \tag{10}$$

If $\tilde{S}_{QQ}(\tau|\beta_1, \beta_2, \psi_T)$ is defined as

$$\tilde{S}_{QQ}(\tau|\beta_1, \beta_2, \psi_T) = \frac{< \psi_T|e^{-\beta_2\hat{H}}\, e^{\tau\hat{H}}\, \hat{Q}^\dagger\, e^{-\tau\hat{H}}\, \hat{Q}\, e^{-\beta_1\hat{H}}|\psi_T >}{< \psi_T|e^{-\beta_2\hat{H}}\, e^{-\beta_1\hat{H}}|\psi_T >} \tag{11}$$

and β_1 and β_2 are chosen sufficiently large so that

$$e^{-\beta_1\hat{H}}|\psi_T > \simeq e^{-\beta_1 E_o} < \psi_o|\psi_T > |\psi_o > \tag{12}$$

and

$$e^{-(\beta_2-\tau)\hat{H}}|\psi_T > \simeq e^{-(\beta_2-\tau)E_o} < \psi_o|\psi_T > |\psi_o >, \tag{13}$$

then

$$\tilde{S}_{QQ}(\tau|\beta_1, \beta_2, \psi_T) \simeq S_{QQ}(\tau). \tag{14}$$

Henceforth, it is assumed that β_1 and β_2 are sufficiently large so that the difference between $\tilde{S}_{QQ}$ and S_{QQ} is exponentially small and can be ignored.

A Monte Carlo algorithm for evaluating Eq. (11) for equal-time operators ($\tau = 0$) has been developed and implemented for Hubbard lattice models [15, 16, 17]. These have Hamiltonia of form $\hat{H} = \hat{H}_o + \hat{H}_I$ where

$$\hat{H}_o = -\sum_{\mathbf{r},\mathbf{r}',\sigma} t_{\mathbf{r},\mathbf{r}'}\, \hat{c}^\dagger_{\mathbf{r},\sigma}\, \hat{c}_{\mathbf{r}',\sigma} \tag{15}$$

and

$$\hat{H}_I = U \sum_{\mathbf{r}} \hat{n}_{\mathbf{r},\uparrow} \hat{n}_{\mathbf{r},\downarrow} \tag{16}$$

describe the kinetic and Coulomb energy terms respectively. The generalization of this algorithm for finite τ values is as follows.

$S_{QQ}(\tau)$ is written as

$$\begin{aligned} S_{QQ}(\tau) &= <\psi_T|(\prod_{i_L=m}^{1} e^{-\Delta\tau\hat{H}})\,\hat{Q}^\dagger\,(\prod_{i_R=m+m'-l+1}^{m+m'} e^{-\Delta\tau\hat{H}}) \\ &\quad \times\hat{Q}\,(\prod_{i_R=1}^{m+m'-l} e^{-\Delta\tau\hat{H}})|\psi_T> \\ &\quad \div <\psi_T|\prod_{i_L=m}^{1} e^{-\Delta\tau\hat{H}} \prod_{i_R=1}^{m+m'} e^{-\Delta\tau\hat{H}}|\psi_T>. \end{aligned} \tag{17}$$

m intervals of size $\Delta\tau$ are used to project ψ_o from ψ_T;

$$m\,\Delta\tau = \beta_1 = \beta_2 - \tau_{max}. \tag{18}$$

Correlation functions are obtained at $m' + 1$ imaginary-time values evenly spaced between $\tau = 0$ and τ_{max};

$$l\,\Delta\tau = \tau, \quad l = 0, 1, \cdots, m'. \tag{19}$$

τ is changed by moving $\hat{Q}$ within the exponentials in Eq. (17).

The Trotter approximation,

$$e^{-\Delta\tau(\hat{H}_o+\hat{H}_I)} = e^{-\Delta\tau\hat{H}_o/2} e^{-\Delta\tau\hat{H}_I} e^{-\Delta\tau\hat{H}_o/2} + \mathcal{O}(\Delta\tau^3), \tag{20}$$

where $\Delta\tau$ is chosen sufficiently small so terms of order $\Delta\tau^3$ can be ignored, separates the interaction and kinetic energy terms. The discrete Hubbard-Stratonovic transformation [18],

$$e^{-\Delta\tau\,U n_{\mathbf{r},\uparrow} n_{\mathbf{r},\downarrow}} = \frac{1}{2} \sum_{\sigma_{\mathbf{r}}=\pm1} e^{2a\sigma(n_{\mathbf{r},\uparrow}-n_{\mathbf{r},\downarrow}) - U\Delta\tau(n_{\mathbf{r},\uparrow}+n_{\mathbf{r},\downarrow})/2}, \tag{21}$$

$$\tanh^2 a = \tanh(\Delta\tau\,U/4), \tag{22}$$

replaces the electron-electron interaction with an electron-Ising field interaction. Eq.(17) becomes

$$S_{QQ}(\tau) = \frac{\sum_{\sigma_{i,\mathbf{r}}} W(\sigma)\, S^{\sigma}_{QQ}(\tau)}{\sum_{\sigma_{i,\mathbf{r}}} W(\sigma)} \tag{23}$$

where

$$S^{\sigma}_{QQ}(\tau) = <\psi_T|(\prod_{i_L=m}^{1} e^{-\Delta\tau\hat{h}(\sigma_{i_L})})\,\hat{Q}^\dagger\,(\prod_{i_R=m+m'-l+1}^{m+m'} e^{-\Delta\tau\hat{h}(\sigma_{i_R})}) \tag{24}$$

$$\times \hat{Q}\left(\prod_{i_R=1}^{m+m'-l} e^{-\Delta\tau \hat{h}(\sigma_{i_R})} \right) |\psi_T> / W(\sigma)$$

and

$$W(\sigma) = <\psi_T| \prod_{i_L=m}^{1} e^{-\Delta\tau \hat{h}(\sigma_{i_L})} \prod_{i_R=1}^{m+m'} e^{-\Delta\tau \hat{h}(\sigma_{i_R})} |\psi_T> . \tag{25}$$

Each configuration of fields, $\sigma_{i,\mathbf{r}}$, (space and/or time indices for σ are not always explicitly given in the above and subsequent expressions) describes a spatially-varying and time-dependent non-interacting fermion system with Hamiltonia $\hat{h}(\sigma_{i,\mathbf{r}})$, but it is generally prohibitive to perform the sum over the large number of field configurations in Eq. (23). However, under certain conditions $W(\sigma)$ is positive definite, or nearly so, and the sum is evaluated with importance sampling.

Numerical evaluation of $W(\sigma)$ and $S^{\sigma}_{QQ}(\tau)$ are obtained as follows. ψ_T is taken to be a Slater determinant of single-electron orbitals,

$$|\psi_T> = \hat{c}_N^{\dagger} \cdots \hat{c}_1^{\dagger}|0>, \tag{26}$$

where $\hat{c}_1^{\dagger}$, for example, creates an electron in orbital $|1>$. Since $\hat{h}(\sigma_{i,\mathbf{r}})$ has no explicit electron-electron interactions the evolved state remains a Slater determinant

$$\prod_{i=1}^{k} e^{-\Delta\tau \hat{h}(\sigma_i)} \hat{c}_N^{\dagger} \cdots \hat{c}_1^{\dagger}|0> = \hat{\tilde{c}}_{N,(1\to k)}^{\dagger} \cdots \hat{\tilde{c}}_{1,(1\to k)}^{\dagger}|0> \tag{27}$$

of evolved orbitals

$$\hat{\tilde{c}}_{n,1\to k}^{\dagger} = \prod_{i=1}^{k} e^{-\Delta\tau \hat{h}(\sigma_i)} \hat{c}_n^{\dagger} \prod_{i=k}^{1} e^{\Delta\tau \hat{h}(\sigma_i)} \tag{28}$$

$$= \sum_{\mathbf{r}} <\mathbf{r}|\tilde{n}_{(1\to k)}> \hat{c}_{\mathbf{r}}^{\dagger} \tag{29}$$

where we have temporarily suppressed electron spin indices. Imaginary-time propagated orbitals conserve neither their norms nor their overlaps. Only the mutually orthogonal components of the orbitals are physically significant so periodic orthogonalization of the N orbitals is required for numerical stability.

After orbitals are evolved through all time steps the weight function becomes,

$$W(\sigma) = <0|\hat{\tilde{c}}_{1,(1\to m)_L} \cdots \hat{\tilde{c}}_{N,(1\to m)_L} \times \hat{\tilde{c}}_{N,(1\to m+m')_R}^{\dagger} \cdots \hat{\tilde{c}}_{1,(1\to m+m')_R}^{\dagger}|0>, \tag{30}$$

whose numerical value is the determinant of an $N \times N$ overlap matrix;

$$W(\sigma) = \text{Det } <\tilde{i}_{(1\to m)_L}|\tilde{j}_{(1\to m+m')_R}> . \tag{31}$$

Evaluation of $S^{\sigma}_{QQ}(\tau)$ is somewhat more complicated. Consider the particular case $\hat{Q} = \hat{c}^{\dagger}_{\mathbf{r}}$, $\hat{Q}^{\dagger} = \hat{c}_{\mathbf{r}'}$ so that

$$S_{QQ} \equiv G^{+}(\mathbf{r}, \tau; \mathbf{r}', \tau = 0) \equiv < \psi_o | e^{\hat{H}\tau}\, \hat{c}_{\mathbf{r}}\, e^{-\hat{H}\tau}\, \hat{c}^{\dagger}_{\mathbf{r}'} | \psi_o > . \tag{32}$$

First ψ_T is propagated $m + m' - l$ steps,

$$\prod_{i_R=1}^{m+m'-l} e^{-\Delta\tau\, \hat{h}(\sigma_{i_R})} |\psi_T > = \hat{\tilde{c}}^{\dagger}_{N,(1\to m+m'-l)_R} \cdots \hat{\tilde{c}}^{\dagger}_{1,(1\to m+m'-l)_R} |0> . \tag{33}$$

Next an electron is placed at $\mathbf{r}'$ and the resulting $N+1$-particle state propagated l additional time steps yielding the evolved state

$$\hat{\tilde{c}}^{\dagger}_{\mathbf{r}',(m+m'-l\to m+m')_R} \hat{\tilde{c}}^{\dagger}_{N,(1\to m+m')_R} \cdots \hat{\tilde{c}}^{\dagger}_{1,(1\to m+m')_R} |0> . \tag{34}$$

Multiplying Eq. (34) by the left-propagated state and the electron annihilation operator for position $\mathbf{r}$ and dividing by $W(\sigma)$ gives

$$\begin{aligned} G^{+\sigma}(\mathbf{r}, \tau; \mathbf{r}', \tau = 0) \;=\; & < 0 | \hat{\tilde{c}}_{1,(1\to m)_L} \cdots \hat{\tilde{c}}_{N,(1\to m)_L}\, \hat{c}_{\mathbf{r}} \\ & \times \hat{\tilde{c}}^{\dagger}_{\mathbf{r}',(m+m'-l\to m+m')_R} \hat{\tilde{c}}^{\dagger}_{N,(1\to m+m')_R} \cdots \\ & \times \hat{\tilde{c}}^{\dagger}_{1,(1\to m+m')_R} |0> / W(\sigma); \end{aligned} \tag{35}$$

whose numerical value is obtained from the determinant of an $(N+1)\times(N+1)$ overlap matrix.

It is unnecessary to evaluate these potentially large determinants for all τ values. Finite τ results are generated from $\tau = 0$ results using the relation

$$\begin{aligned} G^{+\sigma}(\mathbf{r}, \tau; \mathbf{r}', \tau = 0) \;=\; & \sum_{\mathbf{r}''} < \mathbf{r}'' | \tilde{\mathbf{r}}'_{(m+m'-l\to m+m')_R} > \\ & \times G^{+\sigma}(\mathbf{r}, \tau = 0; \mathbf{r}'', \tau = 0). \end{aligned} \tag{36}$$

It is possible to periodically orthogonalize the orbitals for added electrons to the filled orbitals in order to increase the numerical stability of Eq. (36). However, we have not found this to be necessary.

Both electron-addition and electron-removal Green's functions must be evaluated at zero temperature. Annihilation operator propagation is given by

$$\hat{\tilde{c}}_{\mathbf{r}',(m+m'-l\to m+m')_R} \;=\; [\prod_{i_R=m+m'-l}^{m+m'} e^{\Delta\tau \hat{h}(\sigma_{i_R})} c^{\dagger}_{\mathbf{r}'} \prod_{i_R=m+m'}^{m+m'-l} e^{-\Delta\tau \hat{h}(\sigma_{i_R})}]^{\dagger} . \tag{37}$$

The quantity in brackets describes evolution of an electron starting at $\mathbf{r}'$ with $\hat{h}_{-} = -\hat{h}$. It follows that

$$\begin{aligned} G^{-\sigma}(\mathbf{r}, \tau; \mathbf{r}', \tau = 0) \;=\; & \sum_{\mathbf{r}''} < \tilde{\mathbf{r}}'_{-,(m+m'-l\to m+m')_R} | \mathbf{r}'' > \\ & \times [1 - G^{+\sigma}(\mathbf{r}'', \tau = 0; \mathbf{r}, \tau = 0)] \end{aligned} \tag{38}$$

where the "$-$" subscript denotes orbital propagation with $-\hat{h}$.

Density, spin, and current correlation functions are evaluated in terms of $G^{+\sigma}$ and $G^{-\sigma}$ because of the applicability of Wick's theorem for a non-interacting systems. For example, the correlation function

$$S_{\uparrow\uparrow}(\mathbf{r},\tau;\,\mathbf{r}';\tau=0) \equiv <\psi_o|e^{\hat{H}\tau}\,\hat{n}_{\mathbf{r},\uparrow}\,e^{-\hat{H}\tau}\,\hat{n}_{\mathbf{r}',\uparrow}\,|\psi_o> \tag{39}$$

is determined from

$$\begin{aligned} S^{\sigma}_{\uparrow\uparrow}(\mathbf{r},\tau;\mathbf{r}',\tau=0) &= G^{+\sigma}_{\uparrow}(\mathbf{r},\tau;\mathbf{r},\tau)G^{+\sigma}_{\uparrow}(\mathbf{r}',\tau=0;\mathbf{r}',\tau=0) \\ &\quad + G^{+\sigma}_{\uparrow}(\mathbf{r},\tau;\mathbf{r}',\tau=0)G^{-\sigma}_{\uparrow}(\mathbf{r},\tau;\mathbf{r}',\tau=0), \end{aligned} \tag{40}$$

where we have restored electron spin indices. This expression introduces the double-time-displaced Green's function

$$\begin{aligned} G^{+\sigma}_{\uparrow}(\mathbf{r},\tau;\mathbf{r},\tau) &= \sum_{\mathbf{r}',\mathbf{r}''} <\tilde{\mathbf{r}}_{-,(m+m'-l\rightarrow m+m')_R},\uparrow|\mathbf{r}',\uparrow> \\ &\quad\times <\mathbf{r}'',\uparrow|\tilde{\mathbf{r}}_{(m+m'-l\rightarrow m+m')_R},\uparrow> \\ &\quad\times G^{+\sigma}_{\uparrow}(\mathbf{r}',\tau=0;\,\mathbf{r}'',\tau=0), \end{aligned} \tag{41}$$

where both annihilation and creation operators are propagated to τ.

Fig. 1 displays results for $G^{+}(r=0,\tau)$ for the one-dimensional Hubbard model with $U = 4t$ where t is the nearest-neighbor hopping integral in Eq. (15). For equivalent sample sizes (20,000) projector and low temperature ($T = t/16$) grand canonical results agree to within the error bars; these are smaller than the symbol sizes.

The maximum entropy deconvolutions are given in Fig. 1. A qualitatively correct spectral distribution is obtained, though comparison to the exact result demonstrates that QMC data is not informative enough for MEM (or any

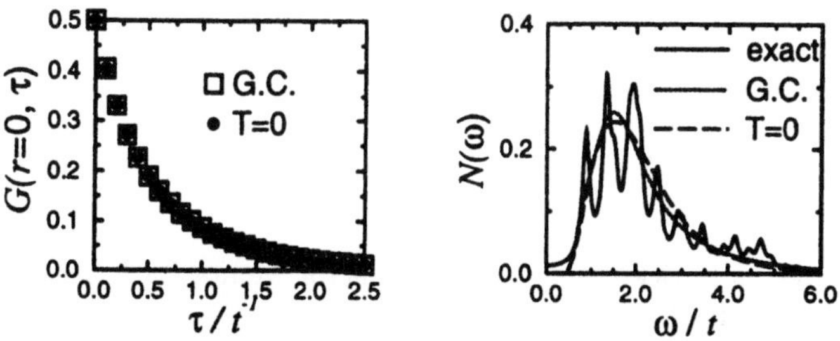

Figure 1: (left) Imaginary-time particle addition correlations from projector and grand-canonical quantum Monte Carlo for the half-filled Hubbard model with 12 sites and $U = 4t$. Error bars are smaller than symbol sizes and are essentially equivalent for the two methods. (right) Analytically continued particle-addition spectrum versus exact diagonalization results. The grand canonical and projector Monte Carlo results are essentially equivalent, providing a qualitatively correct account of the spectral distribution, but broadening the discrete structure.

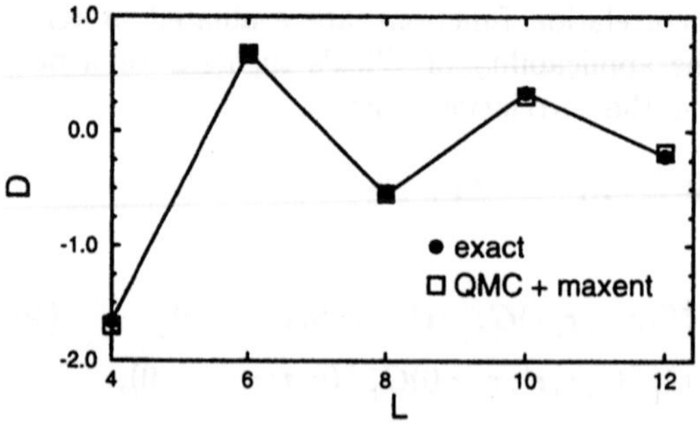

Figure 2: Lattice-size, L, dependence of the Drude weight, D, for $U = 4$ with an average of one electron per site. Quantum Monte Carlo + maximum entropy (QMC + MEM) results agree well with exact values.

other method) to produce all discrete structures in the spectrum. Of course, the advantage of this method is that system sizes can be increased by more than an order of magnitude with respect to exact diagonalization.

Since QMC + MEM spectral distributions are qualitatively correct, subsequent frequency integrals are reasonably accurate. The Drude weight, the $\omega = 0$ part of the optical conductivity, is given by

$$D = - < H_o > -2 \int_0^\infty \frac{S_{jj}(\omega)}{\omega} d\omega. \tag{42}$$

$S_{jj}(\omega)$ is obtained by analytic continuation of the current-current correlation function,

$$S_{jj}(\tau) =< 0|e^{\hat{H}\tau}\, \hat{j}\, e^{-\hat{H}\tau}\, \hat{j}|0 >, \tag{43}$$

where the current operator is

$$\hat{j} = -it \sum_{r,\sigma} (\hat{c}^\dagger_{r,\sigma}\hat{c}_{r+1,\sigma} - \hat{c}^\dagger_{r+1,\sigma}\hat{c}_{r,\sigma}). \tag{44}$$

Fig. 2 displays D versus lattice size, L. The projector Monte Carlo results agree well with exact results.

3. $T = 0$ dynamics for continuous systems.

Diffusion Monte Carlo (DMC) is used for the electron gas model [19], atoms and molecules [20, 21], and periodic atomic lattices [22]. Like the Hubbard model algorithm, DMC uses $\exp(-\hat{H}t)$ as a ground-state projection operator. However, time-dependent particle configurations are sampled rather than Hubbard-Stratonovic fields.

Ground-state projection is carried out in the $3N$-coordinate representation, $\mathbf{R}$, where $|\psi(t)>$ is given by

$$|\psi(t)> = e^{-t\hat{H}}\,|\psi(t=0)> = \int d\mathbf{R}\,\psi(\mathbf{R},t)\,|\mathbf{R}>. \tag{45}$$

The amplitude is determined from the imaginary-time Schrödinger equation

$$-\frac{\partial\psi(\mathbf{R},t)}{\partial t} = [-D\nabla^2 + V(\mathbf{R})\,]\,\psi(\mathbf{R},t) \tag{46}$$

with $D \equiv \hbar^2/2m$. Defining $f(\mathbf{R},t) \equiv \psi_T^*(\mathbf{R})\,\psi(\mathbf{R},t)$, where $\psi_T(\mathbf{R})$ is a trial function, leads to the differential equation,

$$-\frac{\partial f(\mathbf{R},t)}{\partial t} = -D\nabla^2 f(\mathbf{R},t) + E_{local}(\mathbf{R})\,f(\mathbf{R},t) + D\nabla\cdot[f(\mathbf{R},t)\mathbf{F}_Q(\mathbf{R})] \tag{47}$$

with $\mathbf{F}_Q(\mathbf{R}) \equiv \nabla(\ln|\psi_T(\mathbf{R})|^2)$ and $E_{loc}(\mathbf{R}) \equiv <\mathbf{R}|\hat{H}|\psi_T>/\,\psi_T(\mathbf{R})$.

Generally, Eq.(47) cannot be evaluated exactly. However, for small $\Delta\tau$

$$\begin{aligned} G(\mathbf{R}' \to \mathbf{R}, \Delta\tau) &= (4\pi D\,\Delta\tau)^{-3N/2} \\ &\times \exp(-\Delta\tau[E_{loc}(\mathbf{R}) + E_{loc}(\mathbf{R}')]/2) \\ &\times \exp(-[\mathbf{R}' - \mathbf{R} - D\,\Delta\tau\mathbf{F}_Q(\mathbf{R})]^2/(4D\,\Delta\tau)\,) \end{aligned} \tag{48}$$

is a good approximation for the Green's function satisfying

$$f(\mathbf{R}, t+\Delta\tau) = \int d\mathbf{R}'\;G(\mathbf{R}' \to \mathbf{R}, \Delta\tau)\;f(\mathbf{R}',t). \tag{49}$$

Since ground-state projection requires large t values, repeated application of the short-time Green's function is necessary. This yields

$$\begin{aligned} f(\mathbf{R}, n\Delta\tau) &= \int d\mathbf{R}_1 \cdots d\mathbf{R}_n\,G(\mathbf{R}_n \to \mathbf{R}, \Delta\tau)\,\cdots \\ &\quad G(\mathbf{R}_1 \to \mathbf{R}_2, \Delta\tau) f(\mathbf{R}_1, t=0). \end{aligned} \tag{50}$$

For large $n\,\Delta\tau$ the configurational average of $E_{loc}(\mathbf{R})$,

$$< E_{loc}(\mathbf{R}) >= \int d\mathbf{R}\,d\mathbf{R}_1\,\cdots\,d\mathbf{R}_n\,W(\mathbf{R},\cdots\mathbf{R}_n)\,E_{loc}(\mathbf{R}), \tag{51}$$

where

$$W(\mathbf{R},\cdots\mathbf{R}_n) = \frac{f(\mathbf{R}_1, t=0)G(\mathbf{R}_1 \to \mathbf{R}_2, \Delta\tau)\cdots G(\mathbf{R}_n \to \mathbf{R}, \Delta\tau)}{\Omega} \tag{52}$$

and

$$\Omega = \int d\mathbf{R}\cdots d\mathbf{R}_n\;f(\mathbf{R}_1, t=0)\;G(\mathbf{R}_1 \to \mathbf{R}_2, \tau)\;\cdots\;G(\mathbf{R}_n \to \mathbf{R}, \tau), \tag{53}$$

is equal to the ground-state energy. For one- and two-electron systems W is positive definite and stochastic evaluation of Eq. (51) is achieved by sampling particle configuration trajectories $(\mathbf{R}_1, ..., \mathbf{R}_n, \mathbf{R})$.

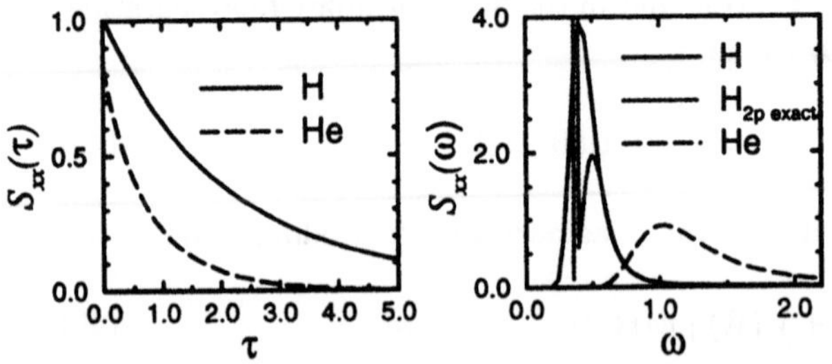

Figure 3: (left) Imaginary-time dipole correlation function and (right) dipole excitation spectrum for hydrogen and helium atoms in atomic units. Similar to Hubbard lattice calculations, qualitatively correct spectral distribution are obtained, but discrete structures are unresolved. When the exact matrix element and frequency for the $1s \rightarrow 2p_x$ transition of hydrogen are included in the MEM default model, structure related to the discrete spectrum begins to appear (dotted line). Improved results are always obtained by including previous knowledge in the default model.

Averages of time-displaced operators,

$$< \hat{Q}(t_1 = n_1\Delta\tau)\, \hat{Q}(t_1 + \tau = n_2\Delta\tau) > \tag{54}$$

$$\equiv \int d\mathbf{R} \cdots d\mathbf{R}_n W(\mathbf{R}, \cdots \mathbf{R}_n)\, Q(\mathbf{R}_{n_1})\, Q(\mathbf{R}_{n_2}),$$

are time-displaced correlation functions. This follows from identities such as

$$\int d\mathbf{R}' G(\mathbf{R}' \rightarrow \mathbf{R}, \Delta\tau) Q(\mathbf{R}')\, f(\mathbf{R}', t) \tag{55}$$

$$= \sum_m e^{-E_m t} < \psi_m|\psi(t=0) > \int d\mathbf{R}' G(\mathbf{R}', \Delta\tau)\psi_T^*(\mathbf{R}')Q(\mathbf{R}')\psi_m(\mathbf{R}') \tag{56}$$

$$= \sum_{m,n} e^{-E_m t} < \psi_m|\psi(t=0) >< \psi_n|\hat{Q}|\psi_m > e^{-E_n\Delta\tau}\psi_T^*(\mathbf{R})\psi_n(\mathbf{R}), \tag{57}$$

that lead to the result

$$< \hat{Q}(t_1 = n_1\Delta\tau)\, \hat{Q}(t_1 + \tau = n_2\Delta\tau) >$$

$$= \sum_m e^{-(E_m - E_o)\tau} | < \psi_m|\hat{Q}|\psi_o > |^2 \tag{58}$$

when $t - t_1 - \tau$ and t_1 are both sufficiently large to project the ground-state.

The left side of Fig. 3 shows dipole correlation functions ($\hat{Q} = \sum_{i=1}^N \hat{x}_i$) for hydrogen and helium atoms. Using a constant default model leads to the hydrogen spectrum represented by the solid line on the right side of Fig. 3.

As with the Hubbard results, the QMC data is not sufficiently informative for MEM to reproduce the discrete spectrum. However, integrated spectral features are again accurate. The QMC + MEM result for the static polarizability is $\chi_{xx}(\omega = 0) = \int \omega^{-1} S_{xx}(\omega) d\omega = 2.246$ comparing well with the exact result of 2.25. It is worth noting that the exact hydrogen ground state is used for ψ_T and $\psi(t = 0)$ in this calculation. While this produces the exact energy via Eq. (51), statistical fluctuations remain in time-dependent correlation functions and an approximate spectrum is produced.

QMC-derived dynamical correlations for atomic helium have been reported earlier [14]. In that work, the real part of the dynamical polarizability was evaluated. In principle, the excitation spectra can then be obtained by a Kramers-Kronig transformation. However, the effective resolution of QMC results makes this transformation numerically unstable. This was the original motivation for applying MEM to QMC data.

The bound-state spectrum of helium begins at $\Delta E_{^1S \to ^1P} \simeq 0.78$ a.u. The discrete low-energy structure is unresolved in our results and merges with excitations above the first ionization threshold at ~ 0.9 a.u. The polarizability obtained with QMC+MEM is 0.657, differing by 3% from the exact value of ~ 0.675.

For systems with more than two electrons, this algorithm must be modified because of the "sign problem": W becomes equally positive and negative in the large t limit. To date, the most effective solution for the sign-problem is the fixed-node approximation where $G(\mathbf{R}' \to \mathbf{R}, \Delta\tau)$ is supplemented with a constraint on electron trajectories that ensures positivity for W [21]. This approximation leads to a variational upper bound for the ground state energy.

The trajectory constraint prevents identification of Eq. (54) as the imaginary-time correlation function. A possible resolution is to use the true Green's function from time slice n_1 to n_2 and the fixed-node Green's function elsewhere. With that approximation Eq. (54) becomes

$$\begin{aligned} &< \hat{Q}(t_1 = n_1\Delta\tau)\, \hat{Q}(t_1 + \tau = n_2\Delta\tau) >_{FN} \qquad (59) \\ &= \sum_m | < \psi_o^{FN} | \hat{Q} | \psi_m > |^2 e^{-E_m\tau} \div \sum_m | < \psi_o^{FN} | \psi_m > |^2 e^{-E_m\tau}. \end{aligned}$$

W becomes negative in this approximation, but its average sign depends only on τ rather than the longer time t needed for both ground state projection and dynamical propagation. We have not yet performed calculations to verify the usefulness of this approach.

Results from *ab initio* calculations can be included in the default model to improve these results. To illustrate this, two MEM deconvolutions are performed for hydrogen and the results shown in Fig. 3. The solid line is the spectrum derived using a constant default model. The dotted line incorporates the exact oscillator strength and energy for the $1s \to 2p_x$ transition in the default model. The dotted line begins to show the gap between the lowest transition and higher transitions. Information from other methods can always be included in these calculations through the default model.

Acknowledgments

We would like to acknowledge useful conversations with F.F. Assaad. Part of this work was supported by a grant from the Deutsche Akadmischer Austauschdienst.

References

[1] R.N. Silver, D.S. Sivia, and J.E. Gubernatis, Phys. Rev. B **41**, 2380 (1990); and in *Quantum Simulations of Condensed Matter Phenomena*, edited by J.E. Gubernatis and J.D. Doll, (World Scientific, Singapore, 1990), p. 340.

[2] J.E. Gubernatis, M. Jarrell, R.N. Silver, and D.S. Sivia, Phys. Rev. B **44**, 6011 (1991).

[3] J.W. Negele and H. Orland, *Quantum Many-Particle Systems*, (Addison-Wesley, Redwood City, 1988), ch. 8.

[4] J. Skilling, in *Maximum Entropy and Bayesian Methods*, ed. J. Skilling (Kluwer Academic Publishers, Dordrecht, 1989), p. 45; in *Maximum Entropy and Bayesian Methods*, ed. P.F. Fougère (Kluwer Academic Publishers, Dordrecht, 1990), p. 341.

[5] S.F. Gull, in *Maximum Entropy and Bayesian Methods*, ed. J. Skilling (Kluwer Academic Publishers, Dordrecht, 1989), p. 53.

[6] S.R. White, Phys. Rev. B **44**, 4670 (1991).

[7] R.N. Silver, D.S. Sivia, J.E. Gubernatis, and M. Jarrell, Phys. Rev. Lett. **65**, 496 (1990).

[8] J. Deisz, M. Jarrell, and D.L. Cox, Phys. Rev. B **42**, 4869 (1990).

[9] For a review of quantum Monte Carlo applications of the maximum entropy method, see M. Jarrell and J.E. Gubernatis, preprint (1995).

[10] R. Preuss, A. Muramatsu, W. von der Linden, P. Dietrich, F.F. Assaad, and W. Hanke, Phys. Rev. Lett **73**, 732 (1994).

[11] J. Deisz, M. Jarrell, and D.L. Cox, Phys. Rev. B **48**, 10227 (1993).

[12] M. Macivić and M. Jarrell, Phys. Rev. Lett **68**, 1770 (1992).

[13] W. von der Linden, Phys. Rep. **220**, 53 (1992).

[14] M. Caffarel, M. Réraf, and C. Pouchan, Phys. Rev. A **47**, 3704 (1993).

[15] S. Sorella, E. Tosatti, S. Baroni, R. Car, and M. Parinello, *Proc. Adriatico Research Conf. Towards the Theoretical Understanding of the High-T_c Superconductors*, eds. S. Lundqvist, E. Tosatti, M. Tosi, and L. Yu (World Scientific, Singapore, 1988).

[16] S. Sorella, S. Baroni, R. Car, and M. Parinello, Europhys. Lett. **8**, 663 (1989).

[17] W. von der Linden, I. Morgenstern, and H. De Raedt, Phys. Rev. B **41**, 4669 (1990).

[18] J.E. Hirsch, Phys. Rev. B **28**, 4059 (1983); Phys. Rev. Lett. **51**, 1900 (1983); Phys. Rev. B **31**, 4403 (1985).

[19] D.M. Ceperly and B.J. Alder, Phys. Rev. Lett. **45**, 566 (1980).

[20] B.J. Alder, D.M. Ceperly, and P.J. Reynolds, J. Phys. Chem. **86**, 1200 (1982).

[21] P.J. Reynolds, D.M. Ceperly, B.J. Alder, and W.A. Lester, Jr., J. Chem. Phys. **77**, 5593 (1982).

[22] X.-P. Li, D.M. Ceperly, and R.M. Martin, Phys. Rev. B **44**, 10929 (1991).

Part III

Contributed Papers

Non-frustrated Ferromagnetic Ising Model on the Quasiperiodic Octagonal Tiling: Finite-Size Behaviour

D. Ledue[1], *D.P. Landau*[2] *and J. Teillet*[1]

[1]LMA URA CNRS 808, Faculté des Sciences, 76821 Mont-Saint-Aignan Cédex, France

[2]Center for Simulational Physics, The University of Georgia, Athens, Georgia 30602, USA

Abstract. The non-frustrated ferromagnetic Ising model on the quasiperiodic octagonal tiling is studied by means of Monte-Carlo simulations. From a finite-size scaling analysis of octagonal tilings with free boundary conditions, the critical temperature is estimated at $kT_c/J = 2.39 \pm 0.01$ and the critical exponents ν, β and γ are in reasonable agreement with previous studies on the Penrose tiling. This strongly suggests that the two-dimensional ferromagnetic Ising model on quasiperiodic tilings and on periodic lattices belong to the same universality class.

1. Introduction

Unlike periodic structures, quasiperiodic tilings, studied by R. Penrose twenty years ago [1], seemed to have a poor interest in physics until the discovery of quasicrystals in 1984 [2]. Since this discovery, one-dimensional (1D), two-dimensional (2D) and three-dimensional (3D) quasiperiodic structures have been widely investigated [3,4]. While the first papers were only concerned with topologic properties of quasiperiodic structures, physical properties of quasiperiodic systems of atoms or spins have been investigated later [5,6]. In particular, numerical investigations about the Ising and the XY model on the Penrose tiling which cannot be solved analytically have been carried out and leaded to results described below. For the frustrated XY model, numerical simulations allowed to determine some frustrated magnetic ground states and to evidence a Kosterlitz-Thouless like transition [7]. For the frustrated antiferromagnetic Ising model, a complex phase diagram has been determined using periodic approximants [6]. For the non-frustrated ferromagnetic Ising model, two Monte-Carlo (MC) investigations using periodic approximants with periodic boundary conditions provided close critical temperatures $kT_c/J = 2.392 \pm 0.004$ and $kT_c/J = 2.401 \pm 0.005$ [8,9]. These studies also evidenced that the non-frustrated ferromagnetic Ising model on the Penrose tiling belong to the same universality class as the ferromagnetic Ising model on 2D periodic lattices.

In this paper, we investigate the static critical behaviour of the non-frustrated ferromagnetic Ising model on the 2D quasiperiodic octagonal tiling by means of MC simulations. Unlike previous works, this study has been done using real finite quasiperiodic tilings with free boundary conditions. Quasiperiodic tilings are defined in Section 2. The models, simulation techniques and finite-size scaling are presented in Section 3. Results are discussed in Section 4 and a summary is given in Section 5.

Springer Proceedings in Physics, Vol. 80
Computer Simulation Studies in Condensed-Matter Physics VIII
Editors: D.P. Landau · K.K. Mon · H.-B. Schüttler

2. Quasiperiodic tilings

Only 2-fold, 3-fold, 4-fold and 6-fold symmetries are consistent with periodicity. So, lattices with other orientational symmetries (5-fold or 8-fold for example) cannot exhibit a translational periodic long-range order. Actually, the periodicity should be replaced by a translational quasiperiodic long-range order.

A quasiperiodic tiling is constructed by the infinite repetition in space of two or more distinct "unit cells" (Fig. 1). The set of the vertices of such a tiling, which is called a quasilattice, is a lattice that exhibits long-range quasiperiodic translational order and long-range orientational order [4] :

- orientational order : each edge of each unit cell is oriented along one of the set of orientational star axes ;
- quasiperiodic translational order : the mass density function of systems made up of atoms located at the vertices of a quasiperiodic tiling is quasiperiodic (the density function can be expressed as a finite sum of periodic functions where at least two periods are incommensurate).

It should be noted that only the invariant transformation belongs to the symmetry group of a quasilattice. Actually, a quasiperiodic tiling (or the corresponding quasilattice QL) is characterized by the set of isometric transformations that leave invariant an infinite part of the vertices. This set of transformations T is called the symmetry pseudo-group G of the quasiperiodic tiling [3] : $T \in G \Leftrightarrow \mathrm{Card}\,(T(QL) \cap QL) = \infty$ (Card is the number of vertices). Then, there is no non trivial translation of the quasiperiodic tiling that leaves it invariant and there is no rotational point symmetry. Moreover, there is no center of symmetry except certain exceptional cases [4].

Two infinite quasiperiodic tilings are locally isomorphic if they can be made to overlap out to any finite distance by a finite translation. This is equivalent to : the two quasiperiodic tilings exhibit the same local environments with the same

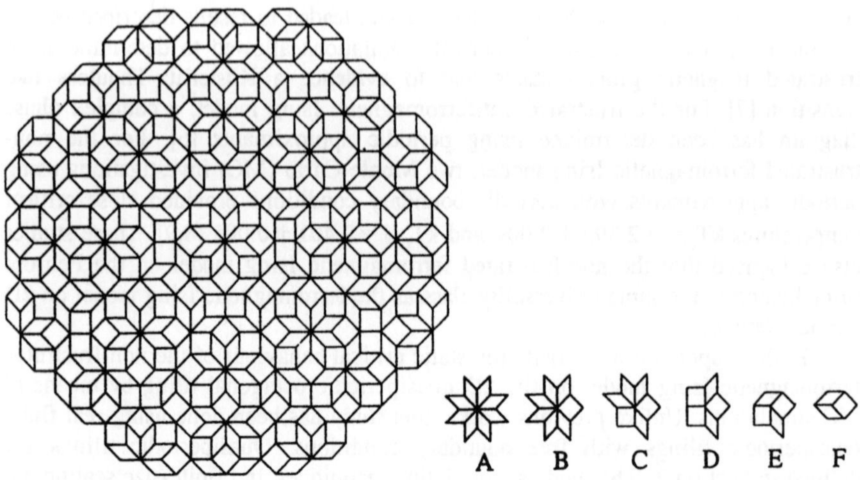

Fig. 1. The octagonal tiling (8-fold orientational symmetry) and the six local environments.

occurrence probabilities. This definition implies that two quasiperiodic tilings which are locally isomorphic exhibit the same orientational symmetry. It can be demonstrated that all the octagonal tilings are locally isomorphic : each octagonal tiling exhibits the six local environments A, B, C, D, E and F (Fig. 1) with the following occurrence probabilities $P(A) \approx 0.02944$, $P(B) \approx 0.01219$, $P(C) \approx 0.05887$, $P(D) \approx 0.14214$, $P(E) \approx 0.34315$ and $P(F) \approx 0.41421$. On the other hand, 5-fold quasiperiodic tilings are not necessary locally isomorphic : 5-fold quasiperiodic tilings belong to several local isomorphism classes. One of these is called the Penrose local isomorphism class [1,4].

Local isomorphism is a very important property because it owns a physical significance. Indeed, two quasiperiodic tilings that are locally isomorphic are equivalent on the average, that is, they exhibit identical macroscopic physical properties. For example, let each vertex of two locally isomorphic quasiperiodic tilings be occupied by a spin. The critical temperatures, the critical exponents and the ground states of the two quasiperiodic systems of interacting spins are identical. Since all the infinite octagonal tilings are locally isomorphic, one has to study only one octagonal tiling. In this paper, we have considered the octagonal tiling with a perfect 8-fold symmetry around its centre (Fig. 1).

Several methods have been described for generating quasiperiodic tilings. The most usual techniques are the generalized dual method (GDM) [4] and the cut and projection method [10]. Let us consider the example of the octagonal tilings generated by the cut and projection method [11].

The 4D integral representation of C_8, the 8-fold cyclic group, decomposes into two inequivalent 2D representations : one operates on $E_{//}$, the "physical space", the other on the orthogonal complementary space $E_{\perp}$. More precisely, $E_{//}$ is defined by the two vectors $\vec{e}_{1,//} = \frac{1}{2}(\sqrt{2}\,\vec{e}_1 + \vec{e}_2 - \vec{e}_4)$ and $\vec{e}_{2,//} = \frac{1}{2}(\vec{e}_2 + \sqrt{2}\,\vec{e}_3 + \vec{e}_4)$ while $E_{\perp}$ is generated by $\vec{e}_{1,\perp} = \frac{1}{2}(\sqrt{2}\,\vec{e}_1 - \vec{e}_2 + \vec{e}_4)$ and $\vec{e}_{2,\perp} = \frac{1}{2}(\vec{e}_2 - \sqrt{2}\,\vec{e}_3 + \vec{e}_4)$, where $(\vec{e}_1, \vec{e}_2, \vec{e}_3, \vec{e}_4)$ is the base of the original lattice Z^4. Then, the octagonal tiling can be obtained by projecting on $E_{//}$ a part of the nodes of Z^4, namely those which have a projection in $E_{\perp}$ falling within an octagonal window [12]. Our octagonal tiling has been generated using in $E_{\perp}$ an octagonal window which is centered on the origin.

3. Background

We have considered the ferromagnetic Ising model with the Hamiltonian :

$$H = -J \sum_{<i,j>} S_i S_j \qquad (J > 0)$$

where the spins S_i, which are located at the vertices of the octagonal tiling, take on the values ±1. The interaction J is along the edges of the unit cells of the tiling.

Numerical simulations were carried out on finite octagonal tilings of size N = 185 up to N = 5497 with free boundary conditions. The procedure is the MC simulated annealing method [13,14]. All simulations were done using the University of Georgia high-performance computer consisting of an IBM ES/9000 model 720 vector supercomputer.

The static critical behaviour of an infinite system may be deduced from the size dependence of the free energy density F [15,16]. Scaling forms for various thermodynamic quantities [17] included the infinite system critical exponents α, β, γ and ν [18] can be obtained from appropriate derivatives of the free energy density. The critical exponent ν can be estimated from the logarithmic derivative ϕ_n of any power of the magnetization [19]. The infinite system critical temperature T_c can be extracted from the location $T_c(L)$ ($L = \sqrt{N}$) of the maximum in ϕ_n, the specific heat C and the "finite system susceptibility" χ' [19] :

$$kT_c(L)/J = kT_c/J + a\, L^{-1/\nu}$$

where a depends on the thermodynamic quantity [20]. The infinite system critical temperature can also be calculated from the fourth-order magnetization cumulant $U(T,L) = 1 - \frac{<m^4>}{3<m^2>^2}$ [21]. The critical exponent β can be deduced from the scaling form of the spontaneous magnetization per spin at T_c while γ can be estimated from the scaling behaviour of the "true susceptibility" at T_c or from the size dependence of the magnitude of the maximum in the "finite system susceptibility".

4. Results

For each size, several runs were performed in order to reduce systematic and statistical errors. The final magnetic quantities are obtained by averaging over the runs.

The estimates of ν deduced from the size dependence of the magnitude of the maximum value of ϕ_1 and ϕ_2 (the logarithmic derivatives of the first and second power of the magnetization) are respectively 0.992 ± 0.009 and 0.995 ± 0.006. Combining these two estimates, we obtain a final value ν = 0.994 ± 0.009 which is in good agreement with the Ising value for 2D periodic lattices (ν = 1).

In Figure 2, we plot the size dependence of $T_c(L)$ extracted from ϕ_1, ϕ_2, χ' and C. The intersections of the four straight lines with the vertical axis provide four values of the infinite octagonal tiling critical temperature : 2.38 ± 0.02, 2.38 ± 0.04, 2.39 ± 0.03, 2.39 ± 0.01. Since the data from the specific heat seem to be more reliable, we estimate the infinite octagonal tiling critical point at $kT_c/J = 2.39 \pm 0.01$.

The temperature dependence of the fourth-order magnetization cumulant U for different sizes is plotted on Figure 3. The infinite tiling critical temperature

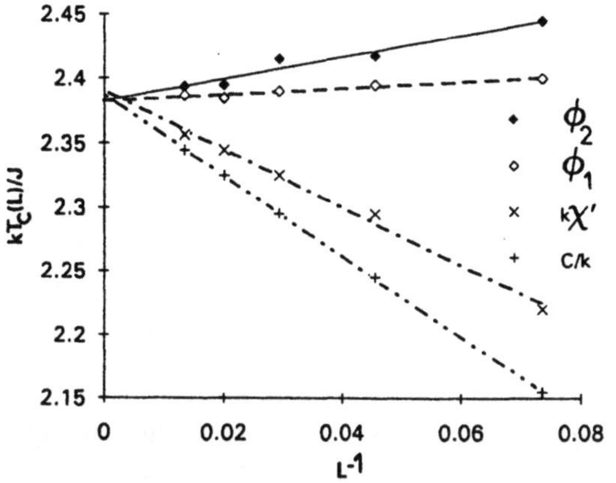

Fig. 2. Size dependence of the location of the maximum in ϕ_1, ϕ_2, C and χ'.

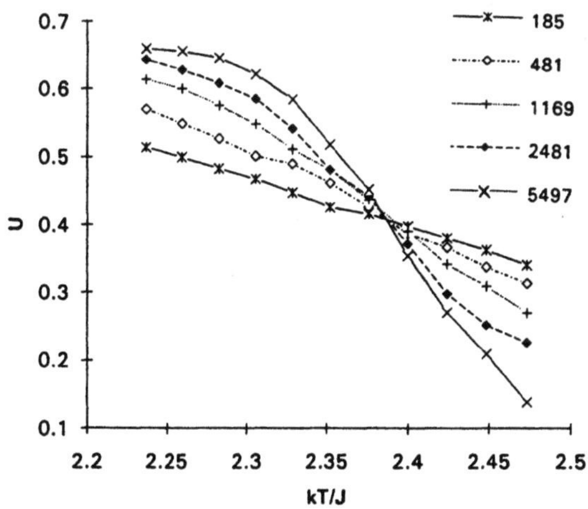

Fig. 3. Temperature dependence of the fourth-order magnetization cumulant for different sizes.

deduced from these curves is $kT_c/J = 2.389 \pm 0.004$ which is in complete agreement with the previous estimate of the critical point.

It should be noted that the infinite octagonal tiling and the infinite Penrose tiling which have the same mean number of interacting neighbours ($\langle z \rangle = 4$) exhibit very close critical temperatures. These critical temperatures are slightly higher than the critical temperature of the square lattice ($kT_c/J = 2.269$) [22].

The plot of the size dependence of the spontaneous magnetization per spin at T_c on a log-log scale [11] provides a slope $\beta/\nu = 0.159 \pm 0.004$ which is slightly higher than the Ising value for 2D periodic lattices ($\beta = 0.125$).

The plots of the size dependence of the "true susceptibility" at T_c and of the size dependence of the maximum in the "finite tiling susceptibility" on a log-log scale [11] provide $\gamma = 1.69 \pm 0.01$ and $\gamma = 1.69 \pm 0.02$. These two estimates lead to a final value $\gamma = 1.69 \pm 0.02$ which is in reasonable agreement with the 2D periodic value ($\gamma = 1.75$).

5. Conclusion

Our investigation of the non-frustrated ferromagnetic Ising model on the octagonal tiling has shown that the finite-size scaling theory describes the size behaviour of finite octagonal tilings with free boundary conditions. Our estimates of the static critical exponents and of the critical point are consistent with previous studies on the Penrose tiling using periodic approximants [8,9]. All these studies strongly suggest that the non-frustrated ferromagnetic Ising model on 2D quasiperiodic tilings belongs to the same universality class as the ferromagnetic Ising model on 2D periodic lattices and that tendency to ferromagnetic ordering is slightly higher in quasiperiodic tilings than in periodic lattices.

In a near future, it is planned to study the static critical behaviour of some frustrated Ising spin systems on the octagonal tiling.

Acknowledgments

The authors wish to thank the North-Atlantic Treaty Organization for providing a fellowship to D. Ledue. D. P. Landau was supported in part by NSF grant N^{O}# DMR-9405018.

References

1. R. Penrose, Bull. Inst. Math and its Appl. 10, 266 (1974)
2. D. Schechtman, I. Blech, D. Gratias and J. W. Cahn, Phys. Rev. Lett. 53, 1951 (1984)
3. D. Gratias, "Du Cristal à l'Amorphe" (Editions de Physique, Paris, 1988)
4. D. Levine and P. J. Steinhardt, Phys. Rev. B 34, 596 (1986)
5. F. Lançon and L. Billard, J. Phys. 49, 249 (1988)
6. J. Oitmaa, M. Aydin and M. J. Johnson, J. Phys. A 23, 4537 (1990)
7. D. Ledue, J. Teillet, J. Carnet and J. Dujardin, J. Non-Cryst. Sol. 153&154, 403 (1993)
8. Y. Okabe and K. Niizeki, J. Phys. Soc. Jpn 57, 16 (1988)
9. E. S. Sorensen, M. V. Jaric and M. Ronchetti, Phys. Rev. B 44, 9271 (1991)
10. A. Katz and M. Duneau, J. Phys. 47, 181 (1986)
11. D. Ledue, D. P. Landau and J. Teillet, to be published in Phys. Rev. B
12. M. Duneau, F. Dunlop, A. Jagannathan and C. Oguey, Mod. Phys. Lett. B 5, 1895 (1991)
13. S. Kirkpatrick, C. D. Gelatt Jr. and M. P. Vecchi, Science 220, 671 (1983)

14. N. Metropolis, A. E. Rosenbluth, M. N. Rosenbluth, A. H. Teller and E. Teller, J. Chem. Phys. **21**, 1087 (1953)
15. M. E. Fisher, "Critical Phenomena", edited by M. S. Green (Academic, New-York, 1971)
16. M. E. Fisher and M. N. Barber, Phys. Rev. Lett. **28**, 1516 (1972)
17. D. P. Landau, Phys. Rev. B **14**, 255 (1976)
18. H. E. Stanley, "Introduction to Phase Transitions and Critical Phenomena" (Clarendon Press, Oxford, 1971)
19. A. M. Ferrenberg and D. P. Landau, Phys. Rev. B **44**, 5081 (1991)
20. K. Chen, A. M. Ferrenberg and D. P. Landau, Phys. Rev. B **48**, 3249 (1993)
21. K. Binder, Z. Phys. B **43**, 119 (1981)
22. L. Onsager, Phys. Rev. **65**, 117 (1944)

A Monte-Carlo Calculation of the Renormalized Coupling Constant for the Three Dimensional Ising Model Using Improved Estimators

G.A. Baker, Jr. and *N. Kawashima**

Theoretical Division, Los Alamos National Laboratory,
University of California, Los Alamos, NM 87545, USA
*On leave from the University of Tokyo

Abstract Improved estimators for the second partial of the susceptibility and the correlation length are developed in the context of the Swendsen-Wang algorithm. These estimators have much improved convergence properties. We report Monte Carlo calculations for the renormalized coupling constant for a series of system sizes and temperatures such that the ratio of the correlation length to the system size is about one-tenth. This ratio is small enough so our results can be expected to approximate the thermodynamic limit within a per cent or better. Our results strongly support hyperscaling.

One of the problems which has troubled the modern renormalization group [1] theory of critical phenomena, particularly the field theory implementation [2], is the validity of those relations between the critical exponents which depend on the spatial dimension (hyperscaling relations). There has been just barely resolved evidence that they are violated. The main test case is the three-dimensional, spin-$\frac{1}{2}$ Ising model. The theoretically most significant indicator in this regard is the renormalized coupling constant, g^*. which is defined as,

$$g^* = \lim_{K\to K_c^-} g(K) = -\lim_{K\to K_c^-} \frac{\dfrac{\partial^2\chi(K)}{\partial H^2}}{\chi^2(K)\xi^d(K)} \tag{1}$$

for zero reduced magnetic field, $H = \beta\mu_B h$, where K_c is the critical value of the inverse temperature $K = \beta J$ with J the exchange energy, χ the susceptibility and ξ is the second moment definition of the correlation length. Field theoretic calculations [3] give $g(K_c, \infty) \approx 23.7$, however Tamayo and Gupta [4] give the results of their Monte Carlo calculations as $g(K_c, \text{large}) \approx 6$.

A number of series analysis of the $g(K,\infty)$ (*e.g.*, [5]) saw a very sharp downturn in the function value as K approached K_c, and suggested that $g(K_c,\infty)$ might actually go to zero, which could imply that the hyperscaling relations failed.

Baker [6] found, in two-dimensions by means of exact calculation on $L \times L$ squares with periodic boundary conditions, the behavior shown in Fig. 1. The main point, for our current purposes, is to notice that the limit as $K \to K_c$ for the infinity system (the top line as computed by series methods) is approached quite well by the finite square results for increasingly larger values of K as L

Springer Proceedings in Physics, Vol. 80
Computer Simulation Studies in Condensed-Matter Physics VIII
Editors: D.P. Landau · K.K. Mon · H.-B. Schüttler

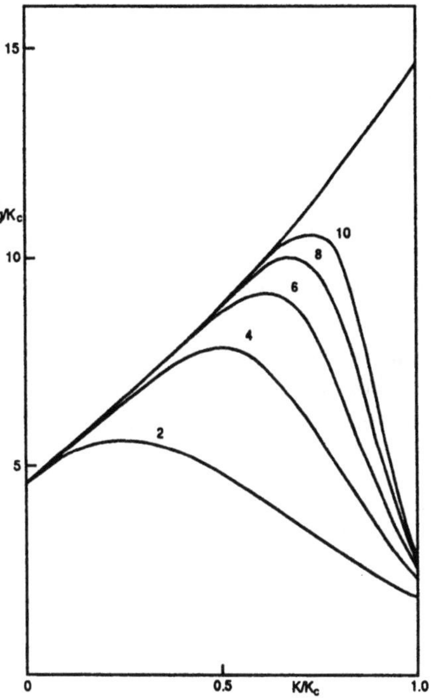

Fig. 1 Plot of gK/K_c *vs.* K. The top line is the series analysis estimate for an infinite system. The labeled curves are for $L \times L$ squares with periodic boundary conditions.

increases, but the limit of $g(K_c, L)$ as $L \to \infty$ has a much smaller value so that $L = \infty$, $K = K_c$ is a point of non-uniform approach. We see this effect in three dimensions in the comparison of the renormalization group values of g^* and the Tamayo-Gupta results mentioned above.

Baker and Erpenbeck [7] have made a study of finite size rounding effects on $g(K, L)$ for the three-dimensional Ising model and found a strong effect. Over the range of $0.25 < \xi_L/L < 0.8$ they found good data collapse for system sizes ranging from $L = 2$ to 100. The resulting curve drops from the vicinity of the renormalization group value for small ξ_L/L to near 3 for the largest values studied. The result of Tamayo and Gupta fits on the curve for $\xi_L/L \approx 0.6$, where ξ_L is the correlation length determined for a system size of L. Baker and Erpenbeck were able to establish in this study that values of $\xi_L/L \geq 0.26$ are too large to use to get the thermodynamic limit. Making ξ/L small is hard to calculate by Monte Carlo because in $\frac{\partial^2 \chi}{\partial H^2} = (\frac{L}{\xi})^d [O(\xi^{\gamma_4/\nu}) - O(\xi^{\gamma_4/\nu})] = O(\xi^{\gamma_4/\nu})$ there is a lot of cancellation! Together all these above mentioned effects make the value of g^* unclear, despite all the previous efforts to obtain it.

In the Swendsen-Wang [8] method one rewrites the partition function,

$$Z = \sum_{\{\sigma_i = \pm 1\}} \exp[K \sum_{\langle ij \rangle} (\sigma_i \sigma_j + 1)]$$
$$= \sum_{\{\sigma_i = \pm 1\}} \sum_{\{b_{ij} = 0,1\}} \prod_{\langle ij \rangle} \left[p \delta_{\sigma_i, \sigma_j} \delta_{b_{ij},1} + \delta_{b_{ij},0} \right], \quad (2)$$

which is a joint distribution over the σ's and the b's, where $p = e^{2K} - 1$. The case $b_{ij} = 1$ has non-zero probability only when the spins σ_i and σ_j are equal, so the sum over spins when the b's are fixed reduces to a sum over cluster spins, σ_c. Thus,

$$Z = \sum_{\{b_{ij}=0,1\}} \sum_{\{\sigma_c=\pm1\}} p^{\sum_{\langle ij\rangle} b_{ij}} = \sum_{\{b_{ij}=0,1\}} p^{\sum_{\langle ij\rangle} b_{ij}} 2^{N_c(\{b_{ij}\})}, \tag{3}$$

where N_c is the number of clusters. The Swendsen-Wang formalism is then, (1) Start with a spin configuration $\{\sigma_i\}$. (2) Set $b_{ij} = 0$ if $\sigma_i \neq \sigma_j$ and set $b_{ij} = 0$ or 1 with relative probabilities $1 : p$ if $\sigma_i = \sigma_j$. (3) Identify the clusters (all the spins which are joined by $b_{ij} = 1$ bonds). (4) Select at random spin values ±1 for each cluster and go back to step (1). Statistics are collected at step (4). Consider the susceptibility,

$$N\chi = \langle\langle \left(\sum_{i=1}^{N} \sigma_i\right)^2 \rangle_b \rangle_\sigma = \langle\langle \left(\sum_c \sigma_c V_c\right)^2 \rangle_{\sigma_c} \rangle_b.$$

where V_c is the number of spins in cluster c. Here the spins (step 4) are independent so $\langle \sigma_c \rangle_{\sigma_c} = 0$, $\langle \sigma_c^2 \rangle_{\sigma_c} = 1$. Thus,

$$N\chi = \langle\langle \sum_c \sum_d \sigma_c \sigma_d V_c V_d \rangle_{\sigma_c} \rangle_b = \langle \sum_c V_c^2 \rangle_b \tag{4}$$

In this method, we effectively sum over all possible choices in step (4) and so get rid of a number of terms which should be exactly zero, without the need to estimate them by Monte Carlo. Using the same methods we can derive,

$$\langle\langle \left(\sum_{i=1}^{N} \sigma_i\right)^4 \rangle_b \rangle_\sigma = \langle\langle \left(\sum_c \sigma_c V_c\right)^4 \rangle_{\sigma_c} \rangle_b = 3\langle \left(\sum_c V_c^2\right)^2 \rangle_b - 2\langle \sum_c V_c^4 \rangle_b, \tag{5}$$

which gives an estimator for

$$-\frac{\partial^2 \chi}{\partial H^2} = (3\langle M^2 \rangle^2 - \langle M^4 \rangle)/N, \tag{6}$$

where $M = \sum_{i=1}^{N} \sigma_i$. We also have derived an improve estimator for the correlation length. To do so, we use the connetion [9] for small wave numbers to the wave-number dependent susceptibility,

$$N\chi(K, \vec{k}) = \left\langle \left| \sum_{\vec{j}} e^{-i\vec{k}\cdot\vec{j}} \sigma_{\vec{j}} \right|^2 \right\rangle, \tag{7}$$

for which in turn we obtain the esitmator,

$$N\chi(K, \vec{k}) = \left\langle \sum_c \left| \sum_{\vec{j} \in c} e^{-i\vec{k}\cdot\vec{j}} \right|^2 \right\rangle_b, \tag{8}$$

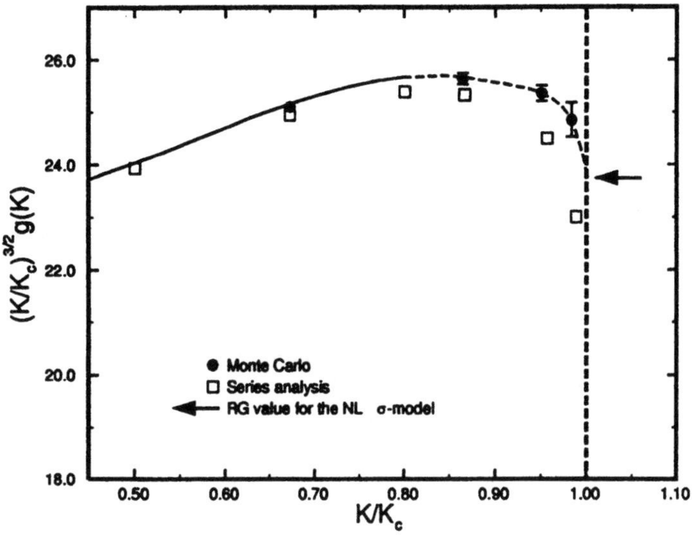

Fig. 2. Computational results for the renormalized coupling constant for the three dimensional Ising model. K_c is taken as the Rosengren conjecture value [3], $\approx 0.22165864\ldots$.

As an example of the effectiveness of these procedures we quote results for the case $K = 0.1916$, $L = 16$. For comparison, we have run the standard Metropolis algothrim, with samples taken every seventh sweep. We ran 70 sweeps for initial equilibration, and then 5,600,000 additional sweeps for measurement, which were coarse grained into 40 bins. We used the Swendsen-Wang algorithm on this same case. Here each of the coarse grained samples consisted of 2,000 sweeps for the initial equilibration and 4,000 sweeps for measurement, making a total of 240,000 sweeps. For the Metropolis simulation, we get $g = 35.5 \pm 2.3$ and $\xi = 1.6070 \pm 0.0039$ while for the Swendsen-Wang simulation we get $g = 31.9 \pm 0.9$ and $\xi = 1.6033 \pm 0.0014$. By way of comparison, for a much longer computation using Swendsen-Wang, we get $g = 31.90 \pm 0.14$ and $\xi = 1.6024 \pm 0.0002$. The results are even more favorable for the improved Swendsen-Wang procedure vs. the Metropolis procedure on larger systems closer to the critical point. In Fig. 2 we give a summary of our computations. The solid line is a new series analysis with more terms. The dashed line is just to guide the eye. It is clear the the old series analysis was dropping too fast, due we believe to an inadequate representation of the behavior close to the critical point. It now appears to us, if the renormalization group results are at least approximately correct, that there is an inverted cusp at the critical point.

References

1. K. G. Wilson, Phys. Rev. B 4, 3174, 3184 (1971).
2. E. Brézin, J.-C. LeGuillou, and J. Zinn-Justin, Phys. Rev. D 15, 1544, 1558 (1977).
3. G. A. Baker, Jr. *Quantitiative Theory of Critical Phenomena*, (Academic Press, Boston, 1990).
4. P. Tamayo and R. Gupta, private communication. We wish to thank these authors for permission to cite this result in advance of publication.
5. G. A. Baker, Jr. and J. M. Kincaid, J. Stat. Phys. 24, 469 (1981).
6. G. A. Baker, Jr., J. Stat. Phys. 77, 955 (1994).
7. G. A. Baker, Jr. and J. J. Erpenbeck, *Computer Simulation Studies in Condensed Matter Physics* 7, eds. D. P. Landau, K.K. Mon and H.-B. Schüttler, ((Springer, New York, 1994) pg. 213.
8. R. H. Swendsen and J. S. Wang, Phys. Rev. Letts. 58, 86 (1987).
9. F. Cooper, B. Freedman, and D. Preston, Nucl. Phys. B210 [FS6], 210 (1991).

Molecular Dynamics Computer Simulation of Cooling Rate Effects in a Lennard-Jones Glass

K. Vollmayr,[†] *W. Kob*[*] *and K. Binder*

Institut für Physik, Universität Mainz,
Staudinger Weg 7, D-55099 Mainz, Germany

Abstract. We present the results of a molecular dynamics computer simulation of a binary Lennard-Jones mixture. We simulate a quench of the system from a liquid state at high temperatures to a glass state at zero temperature by coupling the system to a heat bath that has a temperature that decreases linearly (with slope $-\gamma$) with time. We investigate how the residual density of the system varies as a function of the cooling rate γ and rationalize our results by means of the dependence of the coordination number of the particles on the cooling rate.

Introduction

If a material is cooled rapidly from a high temperature liquid state to a low temperature solid state, crystallization of the substance at its melting point can be avoided and the end product will be an amorphous solid, i.e. a glass. Despite many efforts of experimental and theoretical nature, a good understanding of the relevant mechanism(s) of this so-called glass transition is still lacking. In recent years the so-called mode-coupling theory (MCT)[1] has been used to describe the dynamics of various glass-forming materials. Although the theory seems able to give a correct description of the dynamics of simple glass forming materials, such as colloidal suspensions [2] or a Lennard-Jones system [3], many important materials, such as, e.g., network forming materials like Silica or spin glasses, seem to be inadequately described by MCT. Thus increasing our understanding of the phenomenon of the glass transition remains a challenging problem.

One aspect that makes the investigation of the glass transition difficult, but which is also one of the most outstanding properties of glasses, is the fact that the relaxation times of the material increases to such large values, that the system falls out of equilibrium on the experimental time scale. Hence the glass state produced by a particular process will in general depend on the details of that production process, e.g. on the rate with which the sample was cooled to the glass state. Such a dependence on the cooling rate of the glass transition temperature, or of the density of the glass, has previously been reported, e.g., long ago by Ritland and, more recently, by Brüning *et al* [4].

In this work we present some results of a computer simulation study of a simple glass former, we performed in order to investigate the dependence of various quantities on the cooling rate. The system we deal with has also been investigated in a previous simulation in which it was shown to be a good glass former, i.e. not prone to crystallization even when highly supercooled [3]. Therefore it is interesting to study the glass forming behavior of this system. A more extensive report will be given elsewhere [5].

[†] Supported by Schott Glaswerke, Mainz
Electronic mail: vollmayr@moses.physik.uni-mainz.de

[*] Electronic mail: kob@moses.physik.uni-mainz.de
WWW: http://www.cond-mat.physik.uni-mainz.de/~kob/home_kob.html

Springer Proceedings in Physics, Vol. 80
Computer Simulation Studies in Condensed-Matter Physics VIII
Editors: D.P. Landau · K.K. Mon · H.-B. Schüttler

Model and Details of the Simulation

The model investigated is a binary mixture of Lennard-Jones particles. Both types of particles (type A and type B) have the same mass m. The interaction potential between particles of type α and β ($\alpha, \beta \in \{A, B\}$) is given by $V_{\alpha\beta}(r) = 4\epsilon_{\alpha\beta}\{(\sigma_{\alpha\beta}/r)^{12} - (\sigma_{\alpha\beta}/r)^6\}$. The parameters $\epsilon_{\alpha\beta}$ and $\sigma_{\alpha\beta}$ were chosen as follows: $\epsilon_{AA} = 1.0, \epsilon_{AB} = 1.5, \epsilon_{BB} = 0.5, \sigma_{AA} = 1.0, \sigma_{AB} = 0.8$ and $\sigma_{BB} = 0.88$.

In the following all results are reported in reduced units. The unit of length was chosen to be σ_{AA}, the unit of energy ϵ_{AA} and the unit of time $(m\sigma_{AA}^2/48\epsilon_{AA})^{1/2}$. If the system is assumed to be Argon-like, these units correspond to 3.4Å, 120Kk_B and $3 \cdot 10^{-13}$s. In order to permit a comparison with experiments in which the dependence of the density on the cooling rate is investigated, our molecular dynamics simulations were performed at constant pressure, which was set to 1.0. After an equilibration period at high temperatures ($T = T_0 = 2.0$) the system was cooled to $T = 0$ by coupling each particle to a stochastic heat bath. The temperature T_b of this heat bath was decreased from T_0 to zero linearly with time, i.e. $T_b(t) = T_0 - \gamma t$, where t is the time and γ is the cooling rate. After the system reached the state with $T_b = 0$, i.e. after a time T_0/γ, the resulting configuration of particles was relaxed via a steepest descent procedure to the nearest meta-stable minimum in configuration space. This final configuration was then used for the analysis. In order to improve the statistics, we performed this cooling procedure for each value of γ with ten statistically independent initial configurations. The numbers of A and B particles was 800 and 200, respectively, and the length of the runs for the smallest cooling rate were $8 \cdot 10^6$ MD steps with a step size of 0.02.

Results

In Fig. 1 we show ρ, the density of the system, as a function of T_b, the temperature of the bath, for all values of γ investigated, i.e. $3.1 \cdot 10^{-6} \leq \gamma \leq 1.0 \cdot 10^{-2}$. If the system is assumed to be Argon-like this range of cooling rates corresponds to about $10^9 - 3 \cdot 10^{12}$K/s. Figure 1a shows that, when T_b changes from $T_b = 2.0$ to $T_b = 0$, the density of the system changes by more than a factor of two. We also recognize from this figure that the individual curves do not depend strongly on γ. Only a closer inspection shows that the density $\rho(T_b, \gamma)$ increases systematically with decreasing values of γ. Furthermore it is found that for high values of T_b and not too large values of γ, the curves are identical to within the noise of

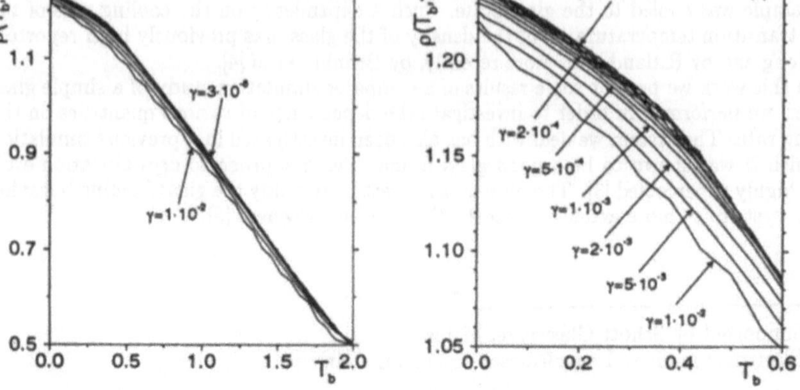

Figure 1: a) Density ρ of the system as a function of T_b, the temperature of the bath, for all cooling rates investigated. b) Same as a) on an expanded scale.

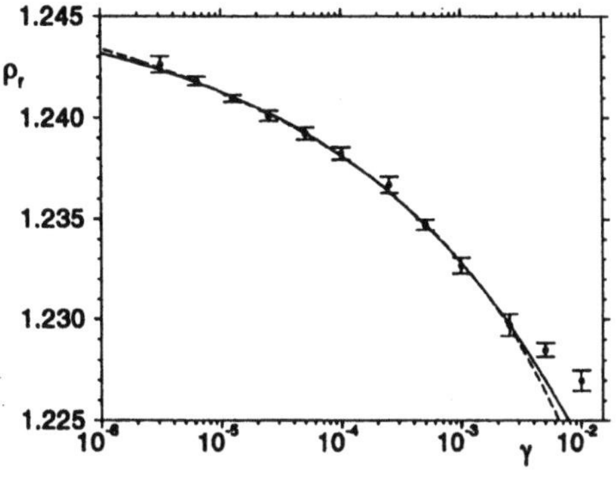

Figure 2: Residual density ρ_r vs. the cooling rate γ. The solid and dashed curves are fits with the functional forms given in Eqs. (1) and (2), respectively.

the data. Thus at these high temperatures and such cooling rates, the system has not yet fallen out of equilibrium and thus the curves follow the *equilibrium* curve $\rho(T_b)$. It is only at low temperatures that the system falls out of equilibrium and the curve for a specific γ will split off from this equilibrium curve. In order to show this effect more clearly we present in Fig. 1b the data at low temperatures on an expanded scale. From this figure it is evident that for temperatures larger than $T_b = 0.5$ the curves for the smallest values of γ still follow the curve $\rho(T_b)$ and only split off at lower temperatures. However, the curves for the larger values of γ have already split off from $\rho(T_b)$ at higher temperatures and are therefore, on this scale, well separated from the curves for the small values of γ. This behavior is very similar to that found in experiments for much smaller cooling rates [4]. Therefore we conclude that, despite the huge difference in cooling rates between our simulation and the experiments (about 10 orders of magnitude), we are able to reproduce at least qualitatively certain aspects of the experiment.

This qualitative similarity is also found in the dependence of the residual density ρ_r on γ, which is shown in Fig. 2. This residual density is the density after the quench to zero temperature and the subsequent steepest descent relaxation of the system. We see that the variation of γ over about three decades gives rise to a variation of ρ_r of about 1%. This figure is close to the one observed in experiments for the same variation of the cooling rate [4], despite the fact that the experimental cooling rates are significantly smaller than the ones investigated here.

Also included in the figure is the result of a fit with a power-law and a logarithmic dependence of ρ_r on γ, i.e.

$$\rho_r(\gamma) = \rho_0 + b_1\gamma^{b_2}. \tag{1}$$

and

$$\rho_r(\gamma) = \rho_0 + a_1(-\log(\gamma))^{a_2} \quad . \tag{2}$$

These functional forms have been motivated from theoretical arguments [6]. We see that for very large cooling rates (the last two points) the dependence of ρ_r on γ apparently changes from the one at small cooling rates in that there seems to be a shoulder in the curve $\rho_r(\gamma)$ close to $\gamma = 2 \cdot 10^{-3}$. We think that at this shoulder the curve makes a crossover from the asymptotic behavior at small cooling rates, to the asymptotic behavior at the infinitely fast cooling rate. Thus we did not include these two last points in the fit. We find that the data at small cooling rates can be fitted equally well with the two functional forms (1) and (2), so with the accuracy of our data it is not possible to determine which functional form is the superior one.

In order to understand the effect of an increasing residual density with decreasing cooling rate from a microscopic point of view, we computed from the final configurations the radial distribution function $g(r)$. In Fig. 3 we show the first peak of $g(r)$ for the BB correlation for all cooling rates investigated. From this figure it is apparent that the height

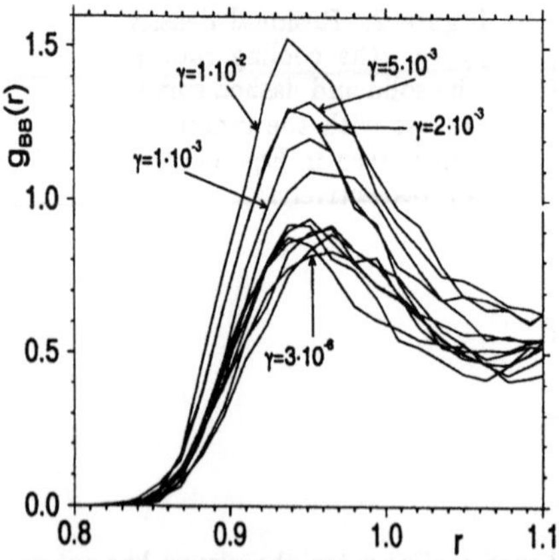

Figure 3: First peak of the radial distribution function $g(r)$ for the BB correlation for all cooling rates investigated.

of this peak decreases when γ decreases. This can be understood by remembering that the strength ϵ of the interaction potential is smaller for the BB interaction than for the AA or AB interaction. Therefore the system will attempt to reduce its energy by trying to avoid two B particles becoming first nearest neighbors. The extend to which it will succeed in this regard increases with the time it is given, i.e. increases with decreasing γ. Thus the number of nearest neighbor pairs of type AB will increase with decreasing γ. Since the σ in the interaction potential is smallest for the AB interaction this will in turn lead to a packing of the particles that is more dense and therefore the residual density will increase with decreasing γ.

In Fig. 4 we plot $P_{BB}(\gamma, z)$, the probability that a B particle has z other B particles as nearest neighbors. The number of nearest neighbors of a particle was defined as the number of particles whose distance to this particle is less than 1.07, the location of the first minimum in $g(r)$ for the BB correlation. One observes from Fig. 4 that the probability that a B particle has no other B particle as its nearest neighbor (curve with $z = 0$), increases with decreasing γ. This is exactly the effect we discussed above. Since, for each value of γ, the sum over all values of z has to add up to unity, this implies that $P_{BB}(\gamma, z)$ has to increase with increasing γ for some of the nonzero z's. This trend is indeed observed for $z = 1, 2$ and 3. Thus this observation is in accordance with our argumentation above for the observed increase of the density with decreasing cooling rate.

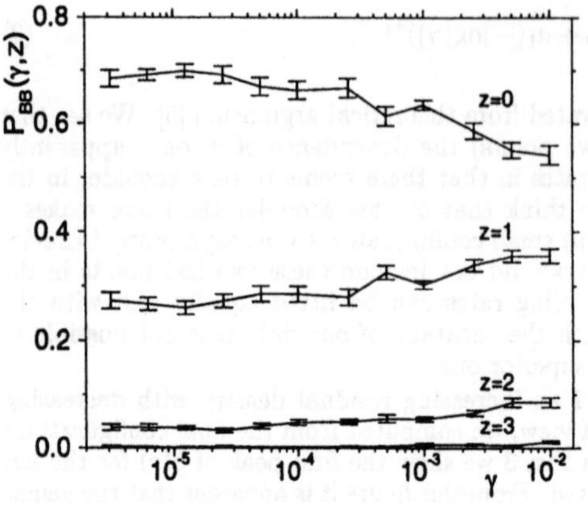

Figure 4: Probability that a B particle has z nearest neighbors of type B.

In conclusion, we have demonstrated that not only global properties (such as the density) but also the local structure of our model glass depends in a characteristic way on the cooling rate, reflecting the tendency of the system to settle down in more stable free energy minima for slower cooling.

Acknowledgements K. V. thanks Schott Glaswerke, Mainz, for financial support. Part of this work was done on the computer facilities of the Regionales Rechenzentrum Kaiserslautern.

References

[1] W. Götze and L. Sjögren, Rep. Prog. Phys. **55**, 241 (1992).

[2] see, e.g., W. van Megen and S. M. Underwood, Phys. Rev. E **49**, 4206 (1994).

[3] W. Kob and H. C. Andersen, Phys. Rev. Lett. **73**, 1376 (1994); W. Kob and H. C. Andersen (preprint).

[4] H, N. Ritland, J. Am. Ceram. Soc. **37**, 370 (1954); R. Brüning and K. Samwer, Phys. Rev. B **46**, 11318 (1992); R. Brüning and M. Sutton, Phys. Rev. B **49**, 3124 (1994).

[5] K. Vollmayr, W. Kob and K. Binder (to be published).

[6] G. S. Grest , C. M. Soukoulis and K. Levin, Phys. Rev. Let. **56**, 1148 (1986).

Coadsorption of Copper and Sulfate on Au(111) Electrodes: Monte-Carlo Simulation of a Lattice-Gas Model

J. Zhang[1], *P.A. Rikvold*[1;2], *Y.-E. Sung*[3], and *A. Wieckowski*[3]

[1]Supercomputer Computations Research Institute, Florida State University, Tallahassee, FL 32306-4052, USA

[2]Department of Physics and Center for Materials Resarch and Technology, Florida State University, Tallahassee, FL 32306-3016, USA

[3]Department of Chemistry, University of Illinois, Urbana, IL 61801, USA

Abstract. We report ground-state calculations and Monte Carlo simulations for a lattice-gas model of the underpotential deposition of copper on Au(111) in sulfate-containing electrolytes. In a potential range of approximately 100~150 mV, this system exhibits a $(\sqrt{3}\times\sqrt{3})$ mixed phase with 2/3 monolayer (ML) copper and 1/3 ML sulfate. Our simulation results agree well with experimental results and with other theoretical work.

1. Introduction

Underpotential deposition (UPD) is a process whereby a monolayer (ML) or less of one metal is electrochemically adsorbed onto another in a range of electrode potentials more positive than those where bulk deposition would occur [1]. The UPD of copper on Au(111) electrodes in sulfate-containing electrolytes has been intensively studied, both experimentally (see detailed discussion in Ref. [2]) and theoretically [3,4]. In cyclic voltammetry (CV) experiments the potential is

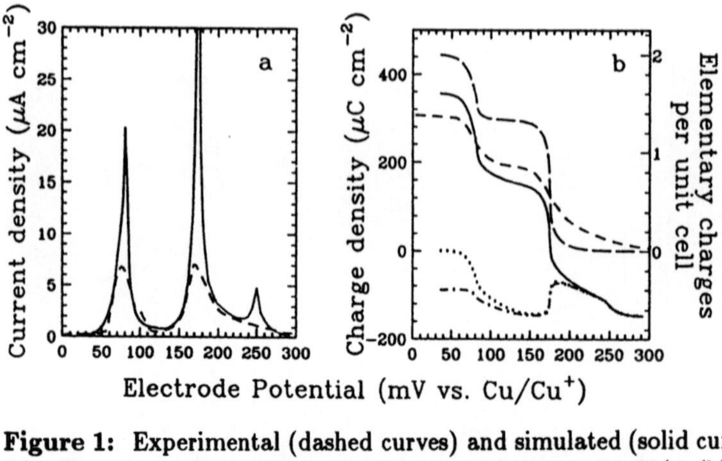

Figure 1: Experimental (dashed curves) and simulated (solid curves) results. (a): CV currents for a positive-going potential scan at 2 mV/s. (b): Integrated charge density. Also shown are $z_C\theta_C$ (long dashed), $z_S\theta_S$ (dotted) and $z_S(\theta_S + \theta_S^{(2)})$ (dotdash).

Springer Proceedings in Physics, Vol. 80
Computer Simulation Studies in Condensed-Matter Physics VIII
Editors: D.P. Landau · K.K. Mon · H.-B. Schüttler

scanned slowly while the current density is recorded. The most striking feature of the CV currents observed with a Au(111) electrode in sulfate-containing electrolyte is the appearance of two peaks, separated by about 100~150 mV, upon the addition of Cu^{2+} ions [5,6]. Typical CV profiles are shown in Fig. 1a [2], together with our simulation results. In the potential range between the peaks, the adsorbate layer is believed to have a $(\sqrt{3}\times\sqrt{3})$ structure consisting of 2/3 ML copper and 1/3 ML sulfate. This picture is based on a variety of both *ex situ* and *in situ* experimental evidence, as discussed in Ref. [2].

2. Model and theoretical methods

Our lattice-gas model for UPD of copper on Au(111) in sulfate-containing electrolyte is a refinement of the Huckaby-Blum model [3,4]. It is defined by a standard three-state lattice-gas Hamiltonian [7,8],

$$\begin{aligned}\mathcal{H}_{\mathrm{LG}} = \sum_l \Big[&-\Phi_{\mathrm{SS}}^{(l)} \sum_{\langle ij\rangle}^{(l)} n_i^{\mathrm{S}} n_j^{\mathrm{S}} - \Phi_{\mathrm{SC}}^{(l)} \sum_{\langle ij\rangle}^{(l)} \left(n_i^{\mathrm{S}} n_j^{\mathrm{C}} + n_i^{\mathrm{C}} n_j^{\mathrm{S}}\right) - \Phi_{\mathrm{CC}}^{(l)} \sum_{\langle ij\rangle}^{(l)} n_i^{\mathrm{C}} n_j^{\mathrm{C}}\Big] \\ &- \Phi_{\mathrm{SS}}^{(t)} \sum_{\Delta} n_i n_j n_k - \bar{\mu}_{\mathrm{S}} \sum_i n_i^{\mathrm{S}} - \bar{\mu}_{\mathrm{C}} \sum_i n_i^{\mathrm{C}} \,.\end{aligned} \tag{1}$$

Here $n_i^{\mathrm{X}} \in \{0,1\}$ is the local occupation variable for species X [X=S (sulfate) or C (copper atom)], and the third adsorption state ("empty" or "solvated") corresponds to $n_i^{\mathrm{S}}=n_i^{\mathrm{C}}=0$. The sums $\sum_{\langle ij\rangle}^{(l)}$, $\sum_{\Delta}$, and $\sum_i$ run over all lth-neighbor bonds, over all equilateral next-nearest-neighbor triangles [9], and over all adsorption sites, respectively, $\Phi_{\mathrm{XY}}^{(l)}$ denotes the effective XY interaction through an lth-neighbor bond, and $\sum_l$ runs over the interaction ranges. The change in electrochemical potential when one X particle is removed from the bulk solution and adsorbed on the surface is $-\bar{\mu}_{\mathrm{X}}$. The sign convention is such that $\Phi_{\mathrm{XY}}^{(l)}>0$ denotes an effective attraction, and $\bar{\mu}_{\mathrm{X}}>0$ denotes a tendency for adsorption in the absence of lateral interactions. The bonds that correspond to finite lateral interaction energies are shown in Fig. 2. For large separations the interactions vanish, and $\Phi_{\mathrm{SS}}^{(1)}$ is an infinite repulsion corresponding to nearest-neighbor sulfate-sulfate exclusion ("hard hexagons" [3,4,10]). We emphasize that the $\Phi_{\mathrm{XY}}^{(l)}$ are *effective* interactions through several channels, including electron-, phonon-, and fluid-mediated mechanisms [7].

The electrochemical potentials in Eq. (1) are (in the weak-solution approximation and here given in molar units) related to the bulk concentrations [X] and the electrode potential E as

$$\bar{\mu}_{\mathrm{X}} = \bar{\mu}_{\mathrm{X}}^0 + RT \ln \frac{[\mathrm{X}]}{[\mathrm{X}]^0} - z_{\mathrm{X}} F E \,. \tag{2}$$

Here z_{X} (X = S,C) are the effective electrovalences of sulfate and copper, R is the molar gas constant, T is the absolute temperature, and F is the Faraday

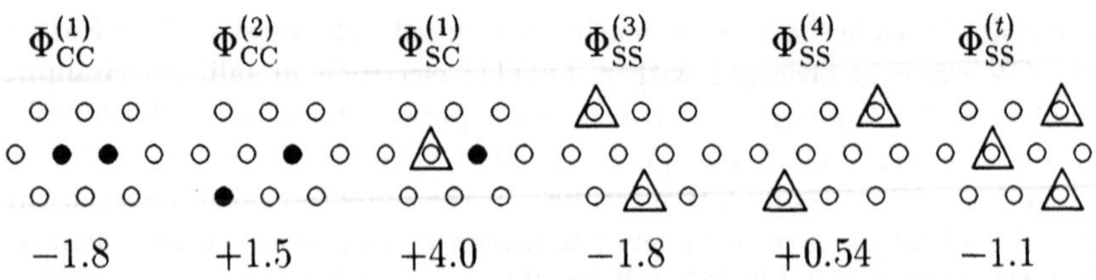

Figure 2: The relative positions of copper (filled circles) and sulfate (triangles) corresponding to the effective interactions used in Eq. (1). The numbers are the corresponding values of $\Phi^{(l)}_{XY}$ used in this work, given in kJ/mol. The interactions are invariant under symmetry operations on the lattice.

constant. The quantities superscripted with a 0 are reference values which contain the local binding energies to the surface. They are generally temperature dependent if the effects of rotational and vibrational modes are considered.

The coverages of sulfate and copper are defined as $\theta_X = N^{-1}\sum n_i^X$, where N is the total number of unit cells in the lattice. However, it is experimentally observed that sulfate remains adsorbed on top of the copper monolayer in the negative-potential region, rather than becoming reduced and entering the solution. A simple estimate of the sulfate coverage in this second layer can be obtained as $\theta_S^{(2)} = \alpha\theta_C(1/3 - \theta_S)$, which allows the difference between the first-layer coverage θ_S and its saturation value of 1/3 to be transferred to the top of the copper layer. The factor α is a phenomenological constant expected to be between zero and one. Since the transfer of sulfate between the gold and copper surfaces does not involve an oxidation/reduction process, the total charge transport per unit cell during the adsorption/desorption process is $q = -e[z_s(\theta_S+\theta_S^{(2)}) + z_C\theta_C]$, where e is the elementary charge. In the absence of diffusion and double-layer effects and in the limit that the potential scan rate $dE/dt \to 0$ [1], the voltametric current i per unit cell of the surface is the time derivative of q. Using differentiation by parts involving the relations between the E and $\bar{\mu}_X$, Eq. (2), as well as the Maxwell relation $\partial\theta_S/\partial\bar{\mu}_C = \partial\theta_C/\partial\bar{\mu}_S$, we find the current density i in terms of the lattice-gas response functions:

$$\begin{aligned} i = eF\Big\{ & z_S^2(1-\alpha\theta_C)\frac{\partial\theta_S}{\partial\bar{\mu}_S}\Big|_{\bar{\mu}_C} + z_C(z_C - 2\alpha z_S\theta_S/3)\frac{\partial\theta_C}{\partial\bar{\mu}_C}\Big|_{\bar{\mu}_S} \\ & + z_S\left(2z_C + \alpha z_S(1/3-\theta_S) - \alpha z_C\theta_C\right)\frac{\partial\theta_S}{\partial\bar{\mu}_C}\Big|_{\bar{\mu}_S}\Big\}\frac{dE}{dt}, \end{aligned} \tag{3}$$

which reduces to its standard form for α=0 [11].

Although the experimental studies are carried out at room temperature, the zero-temperature phase diagram serves as a quite accurate guide to the path in the ($\bar{\mu}_S$,$\bar{\mu}_C$) plane that the isotherms should follow. At constant temperature and pH, two factors influence the path: the adsorbate concentrations in the electrolyte and the electrovalences. As seen from Eq. (2), $\bar{\mu}_S$ and $\bar{\mu}_C$

depend linearly on E, with slopes determined by z_S and z_C, whose values must be determined from experiments. Here we use $z_C=+2$, $z_S=-2$. Thirty-two ordered phases were found by applying the group-theoretical arguments of Landau and Lifshitz [12], nine of which (denoted by $(X\times Y)_{\theta_C}^{\theta_S}$ in Fig. 3) are realized as ground states for interactions in the region of experimental interest. The ground-state energies depend on $\bar{\mu}_S$ and $\bar{\mu}_C$ and the lateral interactions. The repulsive second-neighbor three-particle interaction $\Phi_{SS}^{(t)}$ disfavors the pure sulfate $(\sqrt{3}\times\sqrt{3})_0^{1/3}$ phase, which has not been experimentally observed in this system. For a fixed set of interactions, the zero-temperature phase boundaries are exactly determined by pairwise equating the ground-state energies. In order to easily explore the effects of changing the interactions, a program was developed which numerically determines the zero-temperature phase diagram by scanning $\bar{\mu}_S$ and $\bar{\mu}_C$ and determining the phase of minimum energy [13].

To obtain adsorption isotherms and CV currents at room temperature, we performed Monte Carlo (MC) simulations on a 30×30 triangular lattice, using a heat-bath algorithm [14] with updates at randomly chosen sites. In order to avoid getting stuck in metastable configurations (a problem which is exacerbated by the nearest-neighbor sulfate-sulfate exclusion), we simultaneously updated clusters consisting of two nearest-neighbor sites. Each data point was obtained as an average over 2×10^5 Monte Carlo Steps per spin (MCSS), sampling at intervals of 50 MCSS and discarding the first 4000 MCSS.

3. Numerical results

The zero-temperature phase diagram corresponding to the interactions used in this work is shown in Fig. 3. For large negative $\bar{\mu}_S$ only copper adsorption is possible, and the phase diagram is that of the lattice-gas model corresponding to the triangular-lattice antiferromagnet with next-nearest neighbor ferromagnetic interactions [15]. Similarly, in the limit of large positive $\bar{\mu}_S$ and large negative $\bar{\mu}_C$ the zero-temperature phase is the $(\sqrt{3}\times\sqrt{3})_0^{1/3}$ sulfate phase characteristic of the hard-hexagon model [3,4,10]. The phase diagram for intermediate electrochemical potentials is quite complicated. For $\bar{\mu}_S < -22$ kJ/mol, no sulfate adsorption occurs in the first adlayer, while if $\bar{\mu}_C < -18$ kJ/mol, no copper is adsorbed. The $(\sqrt{3}\times\sqrt{7})_0^{1/5}$ phase corresponds to experimental observations in copper-free systems [16,17]. It is enhanced by the fourth-neighbor sulfate-sulfate attraction, $\Phi_{SS}^{(4)}$, and the $(\sqrt{3}\times\sqrt{3})_0^{1/3}$ phase is disfavored by the second-neighbor repulsive trios, $\Phi_{SS}^{(t)}$. The $(\sqrt{3}\times\sqrt{3})_{2/3}^{1/3}$ mixed-phase region in the upper right-hand part of the diagram is relatively large, due to the nearest-neighbor attraction between copper and sulfate, $\Phi_{SC}^{(1)}$. The isothermal path is chosen such that the distance between points A and B (measured by the electrode potential) equals the peak separation in the CV current shown in Fig. 1a. The narrow strip of $(\sqrt{3}\times\sqrt{7})_{4/5}^{1/5}$ phase lies closely above the $(\sqrt{3}\times\sqrt{3})_{2/3}^{1/3}$ phase and is quite sensitive to the fourth-neighbor attraction, $\Phi_{SS}^{(4)}$, and the second-

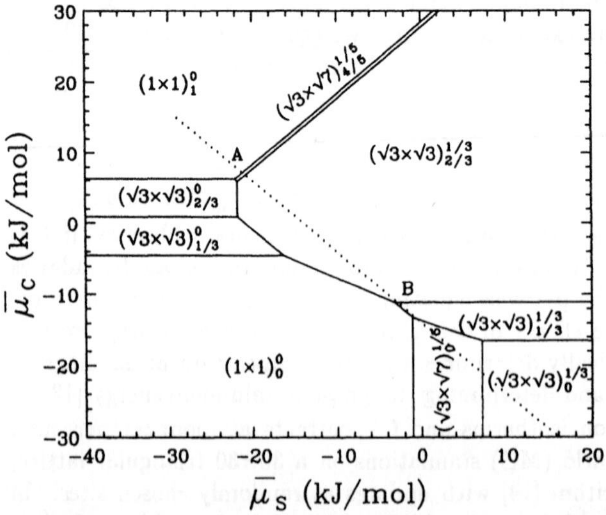

Figure 3: **The zero-temperature phase diagram. Solid lines are phase boundaries and the dotted line is the path along which the isotherms are calculated (positive E towards the lower right). The phases are indicated as $(X \times Y)^{\theta_S}_{\theta_C}$.**

neighbor trio sulfate repulsion, $\Phi^{(t)}_{SS}$. The nearest-neighbor copper repulsion causes the appearance of the $(\sqrt{3}\times\sqrt{3})^{1/3}_{1/3}$ phase.

The potential scan path corresponding to the CV current and adsorption isotherms shown in Fig. 1 is indicated by the dashed line in Fig. 3. With the aid of the ground-state diagram, it is easy to analyse the simulation results. Starting from the positive potential end (lower right in Fig. 3), where $\theta_C \approx 0$, θ_S drops from its saturated hard-hexagon value of 1/3 to approximately 1/5 in the $(\sqrt{3}\times\sqrt{7})^{1/5}_0$ phase region. This is in reasonable agreement with the experimental results for the same electrochemical system in the absence of copper, where the $(\sqrt{3}\times\sqrt{7})^{1/5}_0$ structure has been observed, but not that of $(\sqrt{3}\times\sqrt{3})^{1/3}_0$ [16,17]. As E is scanned in the negative direction, the cations are attracted toward the electrode. Since there are strong effective attractions between the two types of adsorbed particles, the adsorption of copper induces the readsorption of sulfate: they both increase their coverages in a narrow potential range near 175 mV to form the mixed $(\sqrt{3}\times\sqrt{3})^{1/3}_{2/3}$ phase. This is exactly the scenario proposed by Huckaby and Blum [3,4], and it is an example of the enhanced-adsorption phenomenon discussed by Rikvold and Deakin [7]. This phase remains stable in a potential range indicated by the separation of the two CV peaks, until the electrode potential is sufficiently negative that the sulfate adsorption on the gold surface is disfavored in comparison with completion of the copper monolayer. The replacement causes another sharp change in the surface coverages, corresponding to the left CV peak. However, there is experimental evidence that part of the sulfate desorbed from the gold surface is not reduced and dissolved, but rather remains adsorbed in a formally neutral submonolayer on top of the monolayer of copper, with a coverage $\theta_S^{(2)} \approx 0.2$ [6]. This corresponds to α=0.6 in Eq. (3), which was used to obtain the simulated CV current and surface-charge densities shown in Fig. 1. Figure 1b shows a comparison of the simulation results for the charge transfer and the integral of the experimental CV current.

The agreement between the experimental and theoretical results is reasonable, except for large positive E, where the model predicts less copper and more sulfate on the surface than indicated by the experiments. The disagreement between the theoretical and experimental maximum currents may be due to defects on the electrodes used in the experiments.

Acknowledgements

We have enjoyed enlightening conversations with L. Blum. This work was supported by Florida State University through the Supercomputer Computations Research Institute (funded in part by US Department of Energy Contract No. DE-FC05-85-ER25000) and the Center for Materials Research and Technology, and by NSF grant No. DMR-9315969. Work at the University of Illinois was supported by the Frederick Seitz Materials Research Laboratory under US Department of Energy Contract No. DE-AC02-76-ER01198.

References

[1] A.J. Bard and L.R. Faulkner, *Electrochemical Methods* (Wiley, New York, 1985).
[2] J. Zhang, Y.-E. Sung, P.A. Rikvold, and A. Wieckowski, to be submitted to Surf. Sci.
[3] D.A. Huckaby and L. Blum, J. Electroanal. Chem. **315**, 255 (1991).
[4] L. Blum and D.A. Huckaby, J. Electroanal. Chem. **375**, 69 (1994).
[5] M. Zei, G. Qiao, G. Lempfuhl, and D.M. Kolb, Ber. Bunsenges. Phys. Chem. **91**, 3494 (1987).
[6] Z. Shi and J. Lipkowski, J. Electroanal. Chem. **364**, 289 (1993); **365**, 303 (1993).
[7] P.A. Rikvold and Mark R. Deakin, Surf. Sci. **249**, 180 (1991).
[8] P.A. Rikvold, Electrochim. Acta **36**, 1689 (1991), and work cited therein.
[9] S.H. Payne, J. Zhang, and H.J. Kreuzer, Surf. Sci. **264**, 185 (1992).
[10] R.J. Baxter, *Exactly solved models in statistical mechanics* (Academic, London, 1982).
[11] P.A. Rikvold, M. Gamboa-Aldeco, J. Zhang, M. Han, Q. Wang, H.L. Richards, and A. Wieckowski, Surf. Sci., in press.
[12] E. Domany, M. Schick, J.S. Walker, and R.B. Griffiths, Phys. Rev. B **18**, 2209 (1978); E. Domany, and M. Schick, Phys. Rev. B **20**, 3828 (1979); M. Schick, Prog. Surf. Sci. **11**, 245 (1981).

[13] G.M. Buendia, M.A. Novotny, and J. Zhang, in *Computer Simulation Studies in Condensed-Matter Physics VII*, edited by D.P. Landau, K.K. Mon, and H.-B. Schüttler, Springer Proceedings in Physics **78** (Springer, Berlin, 1994).
[14] See, *e.g.*, K. Binder, in *Monte Carlo Methods in Statistical Physics, Second Edition*, edited by K. Binder (Springer, Berlin, 1986).
[15] D.P. Landau, Phys. Rev. B **27**, 5604 (1983).
[16] O.M. Magnussen, J. Hotlos, R.J. Nichols, D.M. Kolb, and R.J. Behm, Phys. Rev. Lett. **64**, 2929 (1990).
[17] G.J. Edens, X. Gao, and M.J. Weaver, J. Electroanal. Chem. **375**, 357 (1994).

Phase Transitions in Multicomponent Widom–Rowlinson Models

J.L. Lebowitz[1], *P. Nielaba*[2] *and L. Šamaj*[3]

[1]Department of Mathematics and Department of Physics, Rutgers University, New Brunswick, NJ 08903, USA
[2]Institut für Physik, KoMa 331, Universität Mainz, D-55099 Mainz, Germany
[3]Courant Institute, N.Y.U., 251 Mercer Street, New York, NY 10012-1185, USA

Abstract. We use Monte Carlo techniques to study the phase diagram of multicomponent Widom-Rowlinson models on a square lattice: there are M species all with the same fugacity z and a nearest neighbor hard core exclusion between unlike particles. For M between two and six there is a direct transition from the gas phase at $z < z_d(M)$ to a demixed phase consisting mostly of one species at $z > z_d(M)$. For $M \geq 7$ there is an intermediate ordered phase in which the even (or odd) sublattice is occupied preferentially by particles chosen at random from any of the species. The existence of such an intermediate phase was proven earlier for $M \geq M_0$, M_0 *very large.* Exact calculations on the Bethe lattice give $M_0 = 4$.

In 1970 Widom and Rowlinson (WR) introduced an ingeniously simple model for the study of phase transitions in fluids [1]. It consists of two species of particles, A and B, in which the only interaction is a hard core between particles of unlike species, i.e. the pair potential $v_{\alpha\beta}(r)$ is infinite if $\alpha \neq \beta$ and $r < R_{AB}$, and is zero otherwise. WR assumed an A-B demixing phase transition to occur in dimensions $D \geq 2$ when the fugacity, $z_A = z_B = z$ is large. A rigorous proof of the existence of a demixing transition was given by Ruelle [2]. His proof, which permits also a smaller hard core $R_{AA} = R_{BB} < (\sqrt{3}/2)R_{AB}$ between like particles was generalized by Lebowitz and Lieb [3] to the case where $v_{AB}(r)$ is large positive but not infinite. An extension of the proof to non-symmetric multicomponent models was made by Bricmont, Kuroda and Lebowitz [4] using Pirogov-Sinai theory. As far as we know, these WR models are the only continuum systems with fixed decaying potentials where one has been able to prove rigorously the existence of phase transitions.

The lattice version of the multicomponent WR model—hard core exclusion between particles of different species on nearest neighbor sites of a simple hyper cubic lattice in D-dimensions—is much easier to handle rigorously. A rather surprising result (at least on first sight) was found by Runnels and Lebowitz [5]. They proved that when the number of components M is larger than some minimum M_0 then the transition from the gas phase at small values of z to the demixed phase at large values of z does not take place directly. Rather there is, at intermediate values of z, $z_c < z < z_d$, an ordered phase in which one of the sublattices (even or odd) is preferentially occupied. The ori-

Springer Proceedings in Physics, Vol. 80
Computer Simulation Studies in Condensed-Matter Physics VIII
Editors: D.P. Landau · K.K. Mon · H.-B. Schüttler

gin of spontaneous symmetry breaking leading to the ordered phase is purely entropic. For z fixed, M large enough it "pays" for the system to occupy just one sublattice without any constraint; there being no interaction between particles on the same sublattice each site can be occupied independently by a particle of any species, i.e. there are M independent choices at each site. This more than compensates for the "loss" of fugacity occasioned by keeping the other sublattice essentially empty.

A natural question is just how big does M_0 have to be to see this ordered phase for $M \geq M_0$. It was shown in [5] that $M_0 < 27^6$, a ridiculously large upper bound. On the other hand a direct computation on the Bethe lattice with q-neighbors [6] gives $M_0 = [q/(q-2)]^2$, which would suggest $M_0 \sim 4$ for the square lattice. The Monte Carlo simulations [6] presented here grew out from a desire to answer this question.

We consider a two dimensional square lattice of size $L \times L$ with periodic boundary conditions. Each of the lattice sites can either be empty or singly occupied by one particle of type I, $I = 1, 2, \cdots, M$. All the components have the same fugacity z and there is an infinite repulsive interaction between particles of different type on nearest neighbor sites. Thus a particle of type I with lattice coordinate (i,j) can only be neighbored by vacancies or particles of type I on nearest neighbor lattice sites (with coordinates $(i, j+1)$, $(i, j-1)$, $(i+1, j)$ or $(i-1, j)$). The interaction potential $\phi_{I,J}(i,j;i',j')$ between a particle of type I with coordinates (i,j) and a particle of type J with coordinates (i',j') is

$$\phi_{I,J}(i,j;i',j') = \begin{cases} \infty \text{ if } \begin{cases} (i,j) \text{ and } (i',j') \text{ are nearest neighbors} \\ \text{and } I \neq J \end{cases} \\ \infty \text{ if } (i,j) = (i',j') \\ 0 \text{ otherwise} \end{cases} \tag{1}$$

We now present the results of the Monte Carlo study [6] of the $M-$ component WR model on the square lattice of size 100×100 with periodic boundary conditions. On an initially empty lattice we deposit particles chosen at random from the M components on lattice sites with probability $z/(z+1)$ respecting the interaction potential (Eq. (1)). We sequentially update the lattice using a checkerboard algorithm resulting in a good vectorization. Typically in a simulation run after an equilibration of 500000 Monte Carlo steps (MCS) we usually update the lattice 500000 times (in the cases $M = 6$ and $z = 3.5$, 4. and 4.5 up to 5000000 times), the configuration of every tenths step is taken for the evaluation of the averages. A typical run with 500000 MCS took about 3 CPU hours on a CRAY–YMP.

Let $M(i,j)$ denote the occupancy of a site, $M(i,j) = 0$ if the lattice site with coordinates (i,j) is empty and $M(i,j) = 1$ otherwise. As observables we took histograms $P_L(\phi_c)$ of the order parameter ϕ_c for the crystal structure and $P_L(\phi_d)$ of the order parameter ϕ_d for the demixed phase in subsystems of size $L \times L$,

$$\phi_c = \frac{1}{L^2} \sum_{i,j=1}^{L} (2M(i,j) - 1)(-1)^{i+j} \tag{2}$$

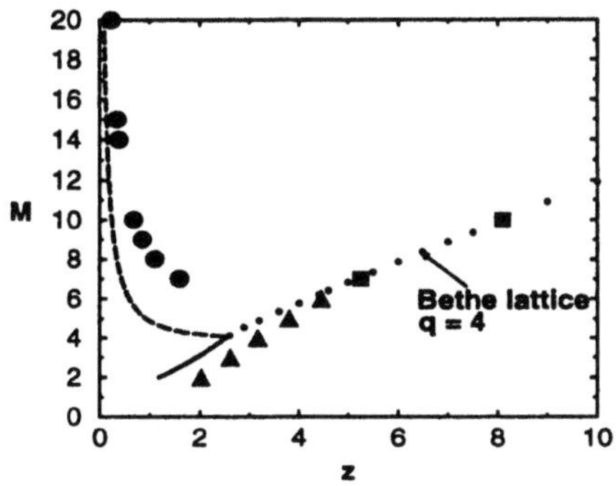

Figure 1: Phase diagram in the M–z plane for a square lattice (MC) and for a Bethe lattice with coordination number $q = 4$. Lines: Exact results for the Bethe lattice for the transition lines from the gas phase to the crystal phase (dashed line), from the gas to the demixed phase (full line) and from the crystal to the demixed phase (dotted line). Symbols for MC: Transition points from the gas phase to the crystal phase (circles), from the gas to the demixed phase (triangles) and from the crystal to the demixed phase (squares).

and

$$\phi_d = \frac{1}{L^2}\text{Max}_I N_L(I) - P/M \tag{3}$$

where Max_I denotes the maximum of the particle numbers $N_L(I)$ of all particle types I in a subsystem of size $L \times L$.

We first discuss the phase transition from the gas to the crystal phase. This phase, in which the density in one of the two sublattices is higher than in the other, is stable for M large enough, due to entropic reasons. Particles preferentially occupy next nearest neighbor positions instead of the nearest neighbor positions due to the "hard square" like nearest neighbor exclusion potential for particles of different type. We locate the transition activity for a given M by finite size scaling techniques [7, 8]; i.e. by looking for intersection points of the fourth order cumulant [8]

$$U_L(\phi_c) = 1 - \frac{<\phi_c^4>_L}{3<\phi_c^2>_L^2}\ , \tag{4}$$

$U_L(\phi_c) = U_{L'}(\phi_c)$ for $L \neq L'$, as a function of the fugacity. $<\phi_c^k>_L := \int \phi_c^k P_L(\phi_c) d\phi_c$ denotes the k–th moment of the order parameter distribution $P_L(\phi_c)$ in subsystems of sizes $L \times L$.

For $M \geq 7$ we find such intersection points indicating the presence of a gas to crystal phase transition, the transition points are shown in Fig. 1. For $M \leq 6$ we were not able to find cumulant intersection points indicating the nonexistence of this transition for $M \leq 6$, for details see Ref. [6].

The phase transitions to the demixed phase with increasing fugacity from either the gas phase or the crystal phase is also found by the analysis of ϕ_d; the transition points are shown in Fig. 1. This phase transition is accompanied by a jump in the density [6] for $M \geq 5$ indicating a first order phase transition for these values of M.

In Fig. 1 we also compare the Monte Carlo results with the results of exact calculations on the Bethe lattice with coordination number $q = 4$. While the topology of the phase diagrams agree, the values of the transition points are different. A rigorous result on the asymptotics is $M/z_d(M) \to 1$ as $M \to \infty$ [6].

Acknowledgements

This research was supported in part by NSF Grant NSF–DMR 92-13424 4-20946. P.N. thanks the DFG for a Heisenberg fellowship, the computations were carried out at the CRAY–YMP of the RHRK Kaiserslautern.

References

[1] B. Widom, J.S. Rowlinson, J. Chem. Phys. **52**, 1670 (1970).

[2] D. Ruelle, Phys. Rev. Lett. **27**, 1040 (1971).

[3] J.L. Lebowitz, E.H. Lieb, Phys. Lett **A 39**, 98 (1972).

[4] J. Bricmont, K. Kuroda, J.L. Lebowitz, Z. Wahrscheinlichkeitstheor. Verw. Geb. **67**, 121 (1984).

[5] L.K. Runnels, J.L. Lebowitz, J. Math. Phys. **15**, 1712 (1974).

[6] J.L. Lebowitz, A. Mazel, P. Nielaba, L. Šamaj, Phys. Rev. **E** (in press).

[7] *Finite Size Scaling and Numerical Simulation*, edited by V. Privman (World Scientific, Singapore, 1990); M.N. Barber, in *Phase Transitions and Critical Phenomena 8*, edited by C. Domb and J.L. Lebowitz (Academic, London, 1983).

[8] K. Binder, Z. Phys. **B 43**, 119 (1981).

Localization Transition in Quantum Percolation

I. Chang[1,2], *Z. Lev*[2], *A.B. Harris*[3], *J. Adler*[2,4], *and A. Aharony*[2]

[1]Department of Physics, Pusan National University, Pusan 609-735, Korea
[2]School of Physics and Astronomy, Tel Aviv University, 69978, Tel Aviv, Israel
[3]University of Pennsylvania, Philadelphia, PA 19104, USA
[4]Department of Physics, Technion-IIT, 32000, Haifa, Israel

Abstract : Connecting perfect one–dimensional leads to sites i and j on the quantum percolation (QP) model, we calculate the transmission coefficient $T_{ij}(E)$ at an energy E near the band center and the averages of $\sum_{ij} T_{ij}$, $\sum_{ij} r_{ij}^2 T_{ij}$ and $\sum_{ij} r_{ij}^4 T_{ij}$ to 10th order in the concentration p. All three series diverge at a finite threshold p_q, with exponents γ, $\gamma + 2\nu$ and $\gamma + 4\nu$. We find $\nu = 0.38 \pm 0.07$in 3 dimensions, differing from "usual" Anderson localization and violating the bound $\nu \geq 2/d$ of Chayes *et al.* Thus, QP belongs to a new universality class.

I. Introduction

The metal-insulator (Anderson localization) transition in the disordered solid has been the subject of intensive studies for a long time.[1,2] The nature of the wavefunction of a single electron in the presence of a random potential determines this transition. Below some critical value of the disorder λ_c the electronic wavefunction is extended, so that the disordered solid shows a metallic behavior. For $\lambda > \lambda_c$ the electronic wavefunction at energy E is localized, so that it shows an insulating behavior: namely, its envelope decays with distance r as $\exp\left[-r/\xi(\lambda, E)\right]$ (apart from power law prefactors), and the localization length ξ diverges as $\xi \sim (\lambda - \lambda_c)^{-\nu}$. The one parameter scaling theory[3] for Anderson localization states that this transition occurs above two dimensions ($2D$), while at and below $2D$ it does not occur for any amount of disorder. On the other hand, it has been a controversial issue whether this transition exists in $2D$.[4,5]

The value of the exponent ν has been the subject of much recent discussion. Particularly, Chayes *et al.*[6] proved (subject to conditions which, though seemingly reasonable, are difficult to verify) that for random systems ν must satisfy the bound $\nu \geq 2/d$. However, they noted that in 3D both the 4–loop ϵ-expansion for $d = 2 + \epsilon$ and some experiments seem to violate this inequality. In this paper we use series expansion for the first accurate numerical determination of ν for the quantum percolation (QP) model,[4,7,8] which is a special variant of the general Anderson model, and find that values of ν, which violate the Chayes *et al.* inequality in both 2D and 3D. Based on our finding we argue that the QP model belongs to a new universality class, different from that of

Springer Proceedings in Physics, Vol. 80
Computer Simulation Studies in Condensed-Matter Physics VIII
Editors: D.P. Landau · K.K. Mon · H.-B. Schüttler

the Anderson localization model. This new universality class may not fulfill the conditions needed for the Chayes *et al.* theorem to apply.

II. Quantum Percolation Model: Generation of Series

The QP model is based on a tight binding Hamiltonian,

$$\mathcal{H} = \sum_i \epsilon_i |i><i| + \sum_{<ij>} t_{ij}(|i><j| + |j><i|), \qquad (1)$$

where $<ij>$ denotes a pair of nearest-neighbor (nn) sites, $|i>$ represents a wave function localized near site i, and we assume a constant on-site energy, ϵ_i, which we arbitrarily take to be zero. The nn hopping matrix element t_{ij} is a random variable which assumes the values 1 or 0 with respective probabilities p and $1-p$. If we identify bonds for which $t_{ij} = 0$ (or 1) as vacant (or occupied), we may view this model as a tight binding model on percolation clusters.

One of the natural and physical measures to probe the localization transition is the average sum over transmission coefficients,[4]

$$T(p,E) = \Big[\sum_{i,j} T_{ij}(E)\Big]_p , \qquad (2)$$

where $T_{ij}(E)$ is the transmission coefficient between sites i and j, the sum is over all pairs of points including $i = j$, and $[\cdots]_p$ represents a configurational average over the t_{ij}'s. To define $T_{ij}(E)$ we attach perfect one-dimensional leads (in which all nearest neighbor t's are unity) to points i and j, insert an incoming wave with energy E on lead i, and calculate the amplitude of the outgoing wave on lead j by the QP model.

As an illustration, we now calculate $T_{ij}(E)$ by solving the Schrödinger equation on a square plaquette (a four bond cluster whose sites are connected in such a way that site 1 is connected to sites 2 and 3, and site 4 is connected to sites 2 and 3). The time independent Schrödinger equation reads

$$E\Psi_k = \sum_\ell t_{k\ell}\Psi_\ell, \qquad (3)$$

where Ψ_k is the amplitude of the wave function near site k, and ℓ runs over the nearest neighbors of k. Assume that a Bloch wave with a wave vector q and an energy $E = 2\cos(qa)$ ($a = 1$ is a lattice constant) enters through a site i, and calculate the amplitude of the outgoing wave at a site j. The tight binding wavefunction on the one dimensional lead is $\Psi_i = e^{inqa} + re^{-inqa}$ ($n = 0, -1, -2, ...$) near site i, and $\Psi_j = te^{inqa}$ ($n = 0, 1, 2, ...$) near site j, where we took $n = 0$ corresponding to sites i and j. Here, r is a reflection constant and t is a transmission constant. Thus at the incoming and outgoing sites the wave function becomes $\Psi_{i=1} = 1 + r$, $\Psi_{j=2} = t$ respectively. Then the set of linear equations (3) for calculating the transmission coefficient $T_{12}(E)$ between sites 1 and 2 are

$$E(1+r) = t + \Psi_3 + e^{-iq} + re^{iq}, \quad Et = \Psi_4 + (1+r) + te^{iq},$$

$$E\Psi_3 = \Psi_4 + (1+r), \quad E\Psi_4 = \Psi_3 + t. \tag{4}$$

Solving these equations gives

$$T_{12}(E) = tt^* = \frac{E^4(4-E^2)}{-E^6+5E^4+2E^2+1}. \tag{5}$$

Given the concentration p of "metallic" bonds, each realization of the system consists of clusters (Γ) of sites interconnected by metallic bonds. For $p \le p_c$ the average $[...]_p$ may be expressed as a sum over the clusters,[9]

$$T(p,E) = \sum_{\Gamma} W_d(\Gamma) p^{n_b(\Gamma)} (1-p)^{n_p(\Gamma)} \sum_{i,j\in\Gamma} T_{ij}(E), \tag{6}$$

where $W_d(\Gamma)$ is the embedding weight (per site) of Γ on the lattice, while $n_b(\Gamma)$ and $n_p(\Gamma)$ are the numbers of bonds in Γ and on its perimeter. If $T_{ij} = 1$ for all i and j on a given cluster, then Eq. (2) reduces to the mean cluster size (or "susceptibility") of classical percolation, which diverges at the classical percolation threshold p_c as $T \sim (p_c - p)^{-\gamma_p}$. In the quantum case, all states are exponentially localized for $p < p_q(E)$, where p_q is the threshold for "quantum percolation" and $p_q \ge p_c$. For $p < p_q$ we expect that in some average sense $[T_{ij}(E)] \sim r_{ij}^{-x} f[r_{ij}/\xi(E)]$, where r_{ij} is the distance between sites i and j, and where $f(x)$ is a scaling function which approaches a constant as $x \to 0$ and decays strongly (e. g. exponentially) for $x \to \infty$. Therefore, $T(p,E) \sim \xi(p,E)^{d-x} \sim (p_q(E)-p)^{-\nu(d-x)} \sim (p_q(E)-p)^{-\gamma}$, with $\gamma = \nu(d-x)$.

Three comments are in order. First, since we take E near the center of the band ($E = 0$), the values of $p_q(E)$ we quote should be close to the critical value p_q^* below which no extended states appear at any energy, studied e. g. in Ref. [8]. Second, since we take E near $E = 0$, one might be worried about the singularities in the density of states which are known to arise at $E = 0$ due to many highly symmetric localized states.[7] To check this, we repeated our series derivation excluding the transmission channels through these states. This was done by expressing T_{ij} in terms of the Green's functions G_{ij}, expanding the latter in terms of the energy eigenstates $|\alpha\rangle$ and excluding the contributions from states with $E_\alpha = 0$. We found that these states do not affect the divergence at p_q. Therefore, we trust that the singularity which we find really indicates a delocalization transition. Third, because all wavefunctions are localized for $p = p_c < p_q(E)$, we do not expect a power–law singularity at $p = p_c$. The absence of a singularity at $p = p_c$ was demonstrated in an exact solution.[10] Also, for the analogous case of the randomly diluted Ising model, no evidence has appeared to suggest a power-law singularity at nonzero temperature at $p = p_c$: The presence of the infinite cluster for $p_c < p < p_q$ does not change the scaling behavior of $T(p, E)$, because the length scale is determined by the localization length and *not* by the percolation correlation length. In any event, our series analysis does not show such a singularity. Therefore, if it exists, its amplitude must be very small or perhaps it is a weak essential singularity.

One way to obtain ξ uses moments of the distances between pairs of sites on the lattice,

$$A_k(p, E) = [\sum_{i,j} r_{ij}^k T_{ij}(E)]_p . \tag{7}$$

Given the above scaling *Ansatz* about the behavior of $T_r(E)$, we expect these moments to diverge near $p_q(E)$ as $A_k(p, E) \sim (p_q(E) - p)^{-\gamma - k\nu}$. Since the sum in Eq. (6) contains polynomials in p, the averages in Eqs. (2) and (7) yield series in p. A calculation to order p^n involves all clusters with up to n bonds.

The series for T require only the topology of the clusters. In contrast, A_k depends on the explicit geometry of each cluster, which requires much more data. To obtain the new series A_2 and A_4, in $d = 3$, we used data files for r_{ij}^2 and r_{ij}^4 on all clusters with $n \leq 10$, which were constructed earlier for the study of the dilute quantum antiferromagnet.[11] For each cluster we calculated $T_{ij}(E = 0.05)$, and then generated the 13 term series for T, and the 10 term series for A_2 and A_4. The series for $E = 0.05$ in 3D are given elsewhere.[12]

III. Analysis of Series

The analysis of the series started by using *d*log Padé analysis[13] for the three series, to identify the rough positions of poles (p_q) and residues ($\gamma + k\nu$). This yielded negative real poles, which were removed by appropriate conformal transformations. We then employed the recently developed efficient three dimensional visualization methods[14] together with the M1 and M2 analysis algorithms,[15] which allow a very accurate determination of the threshold p_q, leading critical exponent (denoted by h in general), and confluent correction to scaling exponent Δ_1 simultaneously. Details of the two methods, M1 and M2 are given in [14] and they are discussed in the article by J. Adler in this volume. In both methods we find the best converging point in the (p_q, h, Δ_1) space, by visualizing 2D plots at different trial p_q values, in perspective.[14] The effectiveness and preciseness of this series analysis method has been demonstrated in recent works.[15,16,17]

From our analyses [12] of the $A_2(p, E = 0.05)$ series for 3D, we found the best convergence at $p_q = 0.36$, with the leading critical exponent $\gamma + 2\nu \approx 1.58$ and the correction to scaling exponent $\Delta_1 \approx 1.05$. We repeated these analyses for T and A_4, and found a central value of $p_q = 0.36$ from all M1 and M2 analyses of all three series, except for M2 on T, which gave $p_q = 0.35$. We quote an overall estimate of $p_q(E = 0.05) = 0.36^{+0.01}_{-0.02}$. This estimate is close to but more precise than earlier series for the inverse participation ratio,[4,8] and other studies.[18,19]

Having identified p_q, we proceeded to deduce overall exponent estimates. The gap exponents between T, A_2, and A_4 should give the estimates for 2ν. Indeed, within the window of $p_q = 0.36^{+0.01}_{-0.02}$ both M1 and M2 gave a constant gap exponent $2\nu = 0.76 \pm 0.14$ as well as $\gamma = 0.82^{+0.10}_{-0.15}$ and $\Delta_1 = 1.05^{+0.20}_{-0.05}$. We find it satisfactory that these three series gave practically the same value of p_q, a constant gap exponent, and moreover Δ_1 close to 1.0. Finally, we performed a self-consistency check for our results stated above, by studying the term-by-term divided series[20,21] of A_4 by A_2. Our M2 analysis of this term-by-term divided series, at $p = 1$, gave $2\nu = 0.75$. The similar M1 analysis yielded $2\nu = 0.74$. Both agree excellently with our direct estimate.

We have also generated and analyzed series for the three quantities discussed above for 2D QP.[22] Consistent with Ref. [4], we found a finite threshold at $p_q = 0.62 \pm 0.01$. In addition, the M1 analysis yielded $\gamma = 1.06 \pm 0.10$, $\nu = 0.50 \pm 0.05$ and $\Delta_1 = 1.20 \pm 0.10$ (M2 did not converge well in this case). We cannot explain why these results differ from those of simulation studies[19].

IV. Discussion and Conclusion

Our main results, $\nu = 0.38 \pm 0.07$ in 3D, and $\nu = 0.50 \pm 0.05$ in 2D, and the existence of a finite threshold in 2D, are surprising for two main reasons: First, our 3D value is quite different from values of ν found for the "usual" Anderson model, which are around 1.35.[23] Assuming the latter values are reliable, we conclude that QP must belong to a different universality class. Our various approximants for the 3D series definitely show no signs of any singularity near $\nu = 1.35$. The existence of a transition in 2D is in contrast to the predictions of the scaling theory [3], which presumably applies to the "usual" localization transition. In some respect, QP involves very strong randomness, as the electron is completely forbidden from hopping onto many sites. Our results indicate that such strong randomness is described by a new set of critical exponents. Second, our results violate the bound $\nu \geq 2/d$ by Chayes *et al.*[6] Although our series are relatively short, and we cannot exclude the possibility of systematic errors, we do not believe that these series can yield a value of ν above 2/3 in 3D (or 1 in 2D). Such values appear only very rarely in our Padé tables, but then require a much larger p_q's (inconsistent with all other results) and different gaps between the 3 series. The internal self-consistency and the agreement of our p_q (at least in 3D) with other methods all lead us to believe that longer series will not yield the large shift needed for obeying the inequality. Attempts to fit the series to essential singularities also failed.[22] The Chayes *et al.* proof is based on a particular finite size scaling correlation length, and they themselves agree that it might not apply to other lengths. Our ξ might indeed differ from theirs. Further, Fisher[24] recently presented a model in which the typical and the average correlation lengths scale differently, and the former violates the Chayes *et al.* inequality. Such a violation can occur when there is a very broad distribution of correlation functions. Returning to our data, we looked at series in which T_{ij} was replaced by a power T_{ij}^{α}, and found that for a wide range of $\alpha > 1$ they all diverged at the same p_q, with an exponent essentially independent of α. This is connected with the fact that for E near 0, most of the typical T_{ij}'s have values very close to 0 or to 1. This results from the fact that for such E, dangling ends tend to generate either full reflection or full transmission. Therefore, the distribution of the T_{ij}'s is extremely broad and this may affect the Chayes *et al.* inequality similarly to what was found in Ref. [24]. The validity of this inequality for QP thus remains in question, and we hope this paper will stimulate more discussion of this issue.

In conclusion, we presented strong evidence that QP belongs to a separate universality class, in which the Chayes *et al.* inequality is violated. It would be useful to test these conclusions using alternative techniques. However, we are not sure if this can be achieved by realistic computer simulations: experience (at least for simple classical spin problems) shows that n-term series yield information equivalent to simulations on lattices of size $(2n)^d$, and calculating the transmission coefficient on such sizes remains a computational challenge. Furthermore, simulations require repeating the calculations for many different realizations, and depend on the quality of the random number generators. Our series contain exact averages over *all* the random configurations.

Acknowledgements: We thank M. E. Fisher, J. T. Chayes, T. C. Lubensky, and Y. Shapir for discussions. We acknowledge support from the U. S. -Israel BSF and the GIF (AA, JA), the NSF grant No. DMR-91-22784 (ABH), the Korea Science and Engineering Foundation through RCDAMP, the Basic Science Research Institute Program (BSRI 93-234, 94-2412) and the University research fund at Pusan National University (IC).

References

1. P. W. Anderson, Phys. Rev. **109**, 1492 (1958).
2. P. A. Lee and T. V. Ramakrishnan, Rev. Mod. Phys. **57**, 287 (1985).
3. E. Abrahams, P. W. Anderson, D. C. Licciardello, and T. V. Ramakrishnan, Phys. Rev. Lett. **42**, 673 (1979).
4. Y. Meir, A. Aharony, and A. B. Harris, Europhys. Lett. **10**, 275 (1989), where a discussion of relevant earlier references is given.
5. C. M. Soukoulis, Q. Li and G. S. Grest, Phys. Rev. B **44**, 4685 (1991).

6. J. T. Chayes, L. Chayes, D. S. Fisher, and T. Spencer, Phys. Rev. Lett. **57**, 2999 (1986).
7. S. Kirkpatrick and T. P. Eggarter, Phys. Rev. B **6**, 3598 (1972).
8. Y. Shapir, A. Aharony, and A. B. Harris, Phys. Rev. Lett. **49**, 486 (1982).
9. J. Adler *et al.*, J. Stat. Phys. **58**, 511 (1990).
10. A. B. Harris, Phys. Rev. Lett. **49**, 296 (1982).
11. C. C. Wan, A. B. Harris, and J. Adler, J. Appl. Phys. **69**, 5191 (1991).
12. I. Chang, Z. Lev, A. B. Harris, J. Adler and A. Aharony, Phys. Rev. Lett. **74**, 2094 (1995).
13. J. L. Gammel, in *Padé Approximants and Their Applications*, ed. B. R. Graves-Morris (Academic Press, New York, N. Y., 1973).
14. J. Adler, I. Chang, and S. Shapira, Int. J. Mod. Phys. C **4**, 1007 (1993).
15. J. Adler, Y. Meir, A. Aharony, A. B. Harris, and L. Klein, J. Stat. Phys. **58**, 511 (1990); J. Adler, M. Moshe, and V. Privman, Phys. Rev. B **26**, 3958 (1982).
16. J. Adler and D. Stauffer, Physica A **181**, 396 (1992).
17. M. Gofman, J. Adler, A. Aharony, A. B. Harris, and M. Schwartz, Phys. Rev. Lett. **71**, 1569 (1993); (E) 2841.
18. T. Odagaki and K. C. Chang, Phys. Rev. B **30**, 1612 (1984); K. C. Chang and T. Odagaki, J. Phys. A **20**, L1027 (1987).
19. C. M. Soukoulis and G. S. Grest, Phys. Rev. B **45**, 7724 (1992).
20. D. L. Hunter and G. A. Baker, Phys. Rev. B **7**, 3346 (1973).
21. Y. Meir, J. Phys. A **20**, L349 (1987).
22. Details will be published elsewhere.
23. e. g., E. Hofstetter and M. Schreiber, Europhys. Lett. **21**, 933 (1993).
24. D. S. Fisher. Phys. Rev. Lett. **69**, 534 (1992).

Vortex Motion in the Two-Dimensional Heisenberg Easy Plane Model

B.V. Costa[1], *D.P. Landau*[2], *J.E.R. Costa*[1], *and K. Chen*[2]

[1]Depto. Física, ICEx, UFMG, CP 702, CEP 30161-790, MG, Brazil
[2]Center for Simulational Physics, University of Georgia,
Athens, GA 30602, USA

Abstract. Using Monte Carlo-Molecular Dynamics simulation properties of the dynamic vortices correlation functions $C(r,t)$ are studied for the two-dimensional classical ferromagnetic easy plane Heisenberg model. Our results suggest that the greatest contribution to $C(r,t)$ comes from a vortex-anti-vortex creation-anihilation process.

1. Introduction

The easy plane anisotropic Heisenberg model in two dimensions provides a prototype for systems which exhibit topological excitations, such as superfluid films, Josephson junction arrays [1] and layered magnets like K_2CuF_4, Rb_2CrCl_4, $CoCl_2$-Graphite intercalation compounds, lipid layers and others [2].
It is well known since the work of Berezinskii [3] and Kosterlitz and Thouless (KT) [4] that the the two-dimensional classical ferromagnetic easy plane Heisenberg model undergoes an infinite order phase transition at the temperature T_{KT}. This phase transition is driven by a vortex-anti-vortex unbinding mechanism. A vortex(anti-vortex) is a topological excitation in which spins on a closed path around the excitation precess by 2π (-2π) in the same direction as shown in Fig. 1. A free vortex is a global excitation while bound pairs are local. (See Fig. 2)
Below T_{KT} vortices and anti-vortices form a condensate of pairs superimposed on spin wave exitations.
The KT theory for the 2D XY model predicts an exponential singularity near T_{KT} with correlation length ξ and magnetic susceptibility χ diverging as

$$\xi \sim e^{b_\xi t^{-\nu}}, \qquad \chi \sim e^{b_\chi t^{-\eta}}, \tag{1}$$

where $t = T - T_{kT}$, with $\nu = \frac{1}{2}$ and $\eta = \frac{1}{4}$. In the disordered phase ($T > T_{KT}$ the vortex density is large and vortex-anti-vortex pairs unbind. In the low temperature phase free vortices are supressed and the dominant excitations are spin waves.
Although static critical properties of the two-dimensional classical ferromagnetic easy plane Heisenberg model are well understood due to extensive studies carried out using a variety of approaches [5], the same is not true about its dynamical behavior.

An ideal gas phenomenology was developed by Mertens et al. [6] which is based on a ballistic approach which was successful in treating one dimensional soliton dynamics in magnetic spin chains [7]. The main ingredient in a such approach is the assumption that a vortex may wander through the lattice obeying a Maxwellian velocity distribution. A vortex passing between the positions $r = 0$ and $r = r_0$

Springer Proceedings in Physics, Vol. 80
Computer Simulation Studies in Condensed-Matter Physics VIII
Editors: D.P. Landau · K.K. Mon · H.-B. Schüttler

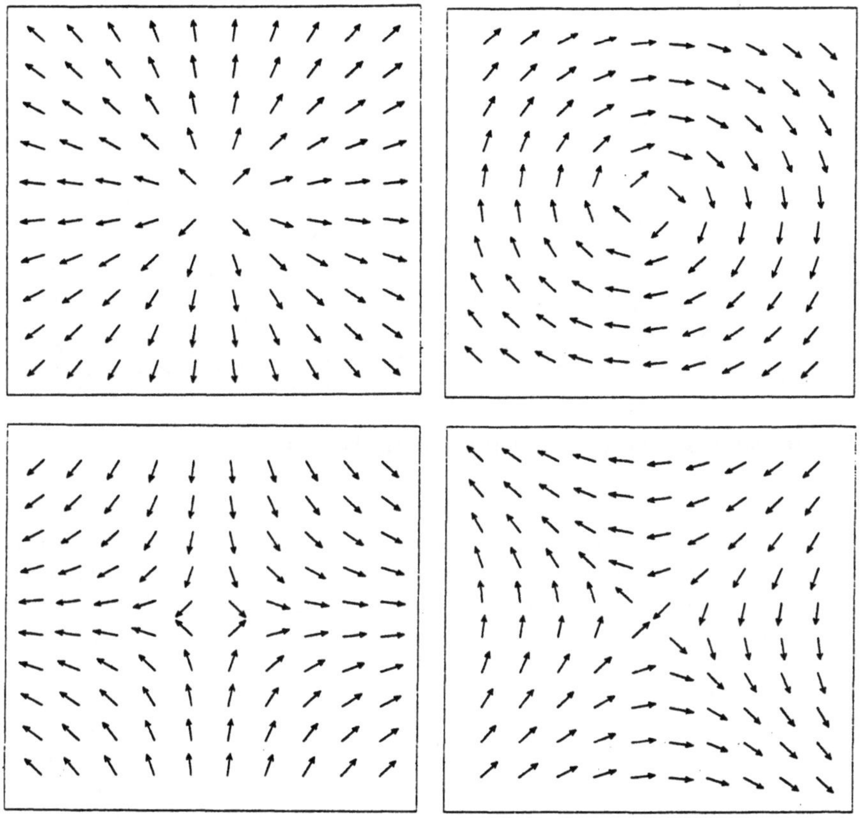

Fig. 1 Examples of vortices **a**, **b** and anti-vortices **c**, **d**.

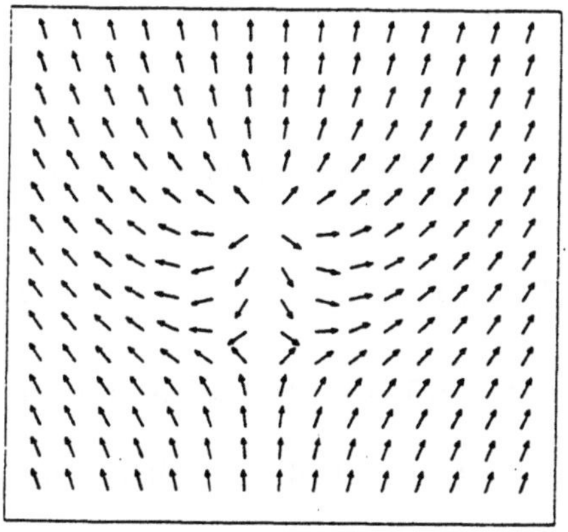

Fig. 2 Vortex-antivortex pair.

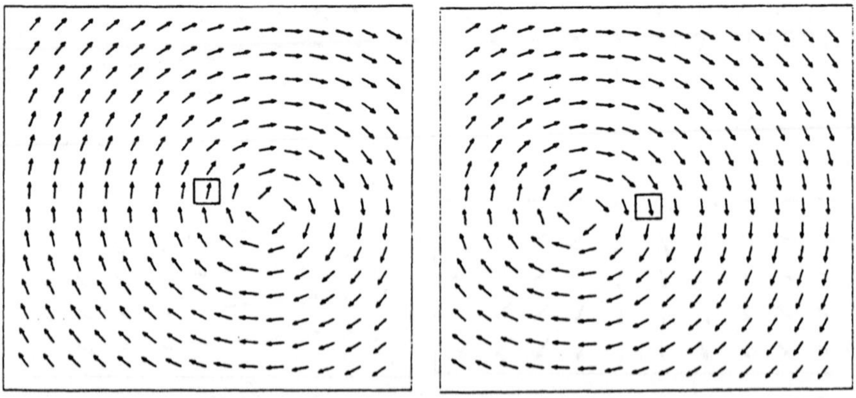

Fig. 3a and b, a moving vortex going through the lattice flips the spin.

flips the spins within that interval (Fig. 3), diminishing the in-plane correlations. Above T_{KT} the authors predicted a Lorentzian squared central peak for the in-plane correlation. Their results are in good qualitative agreement with Monte Carlo - Molecular Dynamics (MC-MD) simulations but fail in the quantitative aspects.

More recently Wiesler et al. [8] tested the Mertens et al. predictions in an extensive neutron scattering experiment in $CoCl_2$ graphite intercalation compound. They were able to measure the scattering intensity as well as the in-plane central peak width Γ_x. The results for the intensity are consistent with both, Lorentzian squared (predicted by Mertens et al.) and simple Lorentzian with comparable values of the χ^2 test. However, the theoretically predicted width Γ_x has a much more rapid increase with temperature than that observed in the experiment.

2. Simulation

In the this work we present some results from MC-MD (or *spin dynamics*) simulations which suggest that in a magnetic lattice vortices have no continuous motion. What our results suggest is that all vortex dynamical behavior is dominated by a creation-annihilation process.

The model which we started with is the XY model defined by the following Hamiltonian

$$H = -J \sum_{<i,j>} (S_i^x S_j^x + S_i^y S_j^y) \quad , \tag{2}$$

where $\vec{S}_i$ is a classical spin variable on the site i defined on the surface of a unit sphere and $J > 0$ is a coupling constant.

In our simulation we have calculated the vortex density fluctuation correlation function C(r,t) defined as

$$C(r,t) = < \Delta\rho(0,0)\Delta\rho(r,t) > / < (\Delta\rho(0,0))^2 >, \tag{3}$$

where $\Delta\rho(r,t) = < \rho(r,t) > - \rho(r,t)$ and ρ is the local vortex density.

Our simulation was performed on a square lattice of side L=44 and periodic boundary conditions. We used a traditional Monte Carlo algorithm by performing 4x10^3 MC steps for each temperature to reach thermodynamic equilibrium and producing 400 initial configurations. To get the dynamical behavior we integrated the equations of motion derived from the Hamiltonian (Eq. 1). We used a time step of $\Delta t = 0.004$ in units of J^{-1}. In this way we got an energy conservation of 10^{-3} percent after 5000 time steps. In order to calculate the correlation C(r,t) we firstly found the vortex (anti-vortex) positions distribution on the lattice for each configuration and time step. Then we divided the lattice into small squares of side l=4. The local vortex (anti-vortex) density $\rho(r,t)$ is just the fraction of vortices inside a square.
With this procedure we have calculated C(r,t) for r= 0, 1, 2 and temperatures from T= .74 to T= .82 in steps of size ΔT= 0.01. Here the temperature T is measured in units of J/k, where k is the Boltzman constant. Some results are shown in Fig. 1 for r= 0 and r= 1. We have adjusted the self-correlation theoretically predicted C(0,t) by using the fitting function

$$C(0,t) = \exp(-Dt^{\nu}). \tag{4}$$

We found that the exponent ν is not sensitive to the temperature. The results are summarized in Table (1). The exponent ν we found ($\simeq .40$) is not compatible with the ballistic approach suggested by Mertens et al, or even with a diffusive behavior. In our simulations we have noted that vortices may not travel through the lattice for more than one or two lattice spacings. Motion in general is followed by an annihilation process between vortices and anti-vortices and then by a creation process. However, we were not able to build up a theory to explain this phenomenon. The local vortex motion is reflected in the $C(r=1,t)$ correlation function shown in Fig. 4. For $r > 1$ we found $C(r,t) = 0$.
Figure 5 shows the *creation-annihilation* rate D versus temperature. The observed values of D were consistent with the divergent form

$$D = A_0 \exp(A_1 \tau^{1/2}) \tag{5}$$

where $\tau = T_{KT}/(T - T_{KT})$ is the reduced temperature, and the right hand side of Eq. (4) is just the static correlation length ξ for the two dimensional planar model powered to some exponent η. In an earlier work Huber [9] found a different form for the diffusion constant D above T_{KT}

$$D \sim (\ln d\xi)^{\frac{1}{2}}, \tag{6}$$

here d is a constant. We believe that such a discrepance lies in the fact that Huber used a continuous approach in his calculation and discrete lattice effects should play an important role.
The value for T_{KT} we have obtained from the fitting is 0.73 which is quite close to the estimate $T_{KT} = 0.725$ obtained from MC simulations [10]. When T approaches T_{KT} from above the rate D grows and the correlation C(0,t) vanishes. The behavior of the rate D suggests that bellow T_{KT} vortex and anti-vortex become a *static* condensate of vortices anti-vortices pairs and the dynamical behavior of the system should be dominated by spin waves excitations.

Table 1 Exponent ν and rate D from the self correlation C(0,t) by using the fitting function, as defined in the text. Temperatures are in units of J/k and the rates D are in units of J^{ν}.

Temperature	Exponent(ν)	Dens. Coef.(D)
0.74	0.40	0.74
0.75	0.40	0.69
0.76	0.41	0.67
0.77	0.42	0.66
0.78	0.40	0.65
0.79	0.40	0.63
0.80	0.41	0.63
0.82	0.40	0.63

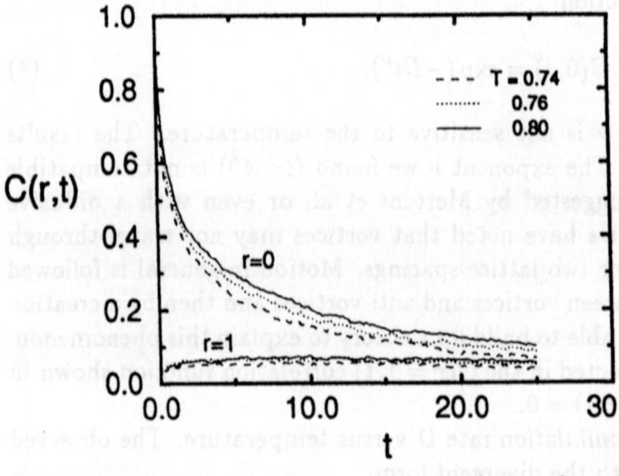

Fig. 4 Vortex density fluctuation correlation function C(r,t) versus time for r= 0, 1 and temperatures T= 0.74, 0.76, 0.80. Temperatures are in units of J/k, r in units of the lattice spacing a and the time is in units of J^{-1}.

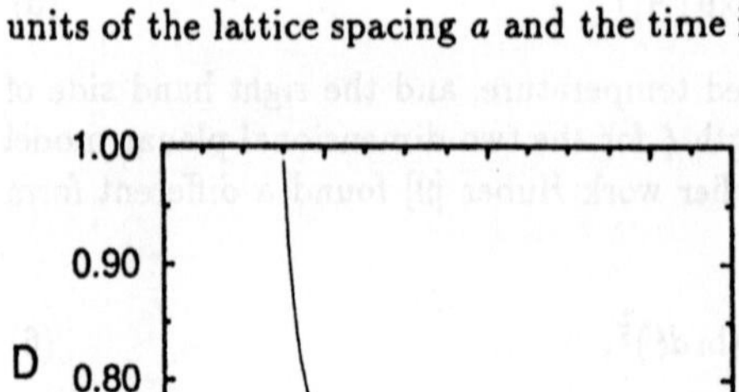

Fig. 5 *Creation-annihilation* rate D versus temperature T. Temperature is in units of J/k and the rate D is in units of J^{ν}.

In conclusion, our simulations suggest that above T_{KT} the vortex dynamics in the XY model is governed by a creation-annihilation process while bellow T_{KT} the vortex anti-vortex condensate has no dynamics.

3. **Acknowledgements** One of us (BVC) would like to thank the warm people of the Center for Simulational Physics at the University of Georgia where part of this work was carried out. This research was partially supported by NSF grant DMR-9405018 and CNPq.

References

1 P. Minnhagen, Rev. Mod. Phys **59**, 1001 (1987);

D. J. Bishop, J. D. Reppy, Phys. Rev. B**22**, 5171 (1980);

D. Mc Qeeney, G. Agnolet and J. D. Reppy, Phys. Rev. Lett. **52**, 1325 (1984);

B. I. Halperin and D. R. Nelson, J. Low. Temp. Phys. **36**, 599 (1979);

R. A. Webb, R. F. Vass, G. Gristein and P. M. Horn, Phys. Rev.

Lett. **51**, 690 (1983).

2 M. Pomerantz, Surf. Sci. **142**, 556 (1984);

D. I. Head, B. H. Blott and D. Melville, J. Phys. C **8**, 1649 (1988).

3 V. L. Berezinskii, Zh. Eksp. Teo. Fiz. **61**, 1144 (1971).

4 J. M. Kosterlitz and D. J. Thouless, J. Phys. C **6**, 1181 (1973).

5 Rajan Gupta, Jerry DeLapp and George G. Batrouni, Phys. Rev. Lett. **61**, 1996 (1988).

6 F. G. Mertens, A. R. Bishop, G. M. Wysin and C. Kawabata, Phys. Rev. Lett. **59**, 117 (1987); Phys. Rev. B **39**, 591 (1989).

7 J. A. Krumhansl and J. R. Schrieffer, Rev. Rev. B **11**, 3535 (1975);

H. J. Mikeska, J. Phys. C: Solid St. Phys. **11**, L29 (1978);

8 D. G. Wiesler, H. Zadel and S. M. Shaphiro, Z. Phys. B **93**, 277. (1994).

9 D.L.Huber, Phys. Rev. B, **26**, 3758 (1982)

10 H.G.Evertz and D.P.Landau, to appear in *Computer Simulation Studies in Condensed Matter Physics* VIII, Eds. D.P.Landau, K.K.Mon and H.B.Schuettler (Springer Verlag 1995).

Thoughts on Interactive Simulation

D.C. Rapaport

Physics Department, Bar-Ilan University, Ramat-Gan, Israel

We briefly outline several computer demonstrations involving interactive computer graphics shown at the Workshop and discuss some of the general principles involved in designing this kind of software.

1. Introduction

Computer simulation plays an ever-increasing role in both the exploratory and didactic aspects of science. The potentially rich environments that can be created within the computer not only provide research scientists with the ability to examine complex phenomena but also enrich the learning process by making similar capabilities available for educational use. There can be little doubt that with the onset of the information age both the practice and teaching of science are undergoing significant change.

From their first appearance computers have played a central role in physics by allowing the simulation of processes that are not otherwise theoretically tractable. This is especially true in statistical physics where the models studied are often simple to formulate but, with just a few notable exceptions, impossible to solve.

For many years, serious work in simulational physics was confined to mainframe computers (indeed these were the only computers available), but with the appearance of minicomputers, followed by personal computers and workstations whose capabilities grow from year to year, the cost of entry into this exciting field has been dramatically reduced. Supercomputer computations of little more than a decade ago can now be carried out on a personal machine – a workstation or high-performance PC – in a much friendlier working environment than that provided by what is now regarded as an archaeological relic (to be found in computer museums). There are now a wide variety of tools to aid software development, and the personal machine reduces the perpetual struggle for computational resources on a public computer because it guarantees a certain minimal (but substantial) level of throughput.

The advent of personal computing has brought additional benefits, in particular the possibility of using interactive graphics as an integral part of a com-

Springer Proceedings in Physics, Vol. 80
Computer Simulation Studies in Condensed-Matter Physics VIII
Editors: D.P. Landau · K.K. Mon · H.-B. Schüttler

putation. The ability to visualize the phenomena under study brings a new level of understanding to the work, because the results can now be examined in ways for which no alternatives exist. The interactivity facilitates exploration: the now available computational power means that results no longer emerge after unbearably long periods of time; in many cases it is now possible to obtain meaningful results and then change certain parameter settings and repeat the operation over and over again. Such an approach engenders a far more thorough understanding of the system under investigation.

In this short paper we outline a number of problems of interest to the statistical mechanics community that benefit from this approach (including those demonstrated during the Workshop) and discuss some of the design considerations involved in developing software of this type.

2. Visualization

The kind of visualization appropriate to a particular problem varies widely and a very small number of typical examples of problems that appeal to statistical physicists will have to suffice here. The amount of computation required, the kinds of interactivity that must be made available to the user, and the complexity of the graphics needed to display the details of the behavior vary widely between the examples. A written description is of course not a substitute for an interactive graphical demonstration, but that is all the printed medium permits.

a) In the case of an Ising model studied by the Monte Carlo method it is possible to examine the individual spins and the clusters they form. While the values of bulk thermodynamic and magnetic properties, as well as spin correlation functions, provide quantitative information of what occurs near the critical point, it is the direct observation of the changing nature of the spin clusters as the temperature is altered that provides a fuller appreciation of the underlying fluctuations that drive the transition. Another problem in which clusters play a key role is percolation, also studied using a Monte Carlo approach, and here again the ability to observe the actual clusters aids in understanding the behavior. In general, where the choice exists, it is preferable to focus on 2D problems because of the visual complexity of the 3D case.

b) Molecular dynamics simulations of fluids provide another example of the value of visualization. In the case of a model monatomic fluid, in which the atoms interact through a Lennard-Jones potential, it is instructive to contrast the atomic trajectories in the fluid and solid states, as well as how the atoms form an ordered solid state subject to lattice imperfections and misaligned domains. For molecules with internal structure – such as rigid linear molecules – the orientational ordering that can occur in the fluid state reminiscent of a nematic liquid crystal can also be examined.

c) Studies of vector spin models, again using molecular dynamics methods, benefit similarly. Here it is possible to observe domains of ordered spins, to detect small clusters of spins that fail to equilibrate for long periods of time, and to see spin waves and 'vortices'. While statistical mechanics addresses equilibrium averages, visualization brings to light the actual behavior, and helps in understanding why, for example, a system may not reach equilibrium as rapidly as might be expected.

d) One of the principal motives for studying fractals is the richness of the visual information they attempt to condense – hence the concept of fractal 'forgery'. Fractals can be used to generate landscapes and other scenes from nature to varying degrees of visual precision based on surprisingly simple recipes. Interactively generating a series of not entirely unrealistic mountainous scenes serves as an interesting exercise; here the large repertoire of techniques used in creating computer imagery (such as illumination and shading) can be brought into play to increase the realism of the scene. The production of deterministic fractal objects such as sponges, as well as the extremely complex Mandelbrot set, are other examples where graphics plays an indispensable role.

e) Aggregation and growth phenomena are examples of fields that developed only after computer graphics became widely available (as is also true for the entire field of fractals). Diffusion-limited aggregation is a well-known means for growing highly ramified cluster-like objects encountered widely in nature; watching the growth process as it occurs is a fascinating experience. Not only is the shape of the growing cluster of interest, but also the equipotential curves that surround it – assuming the cluster to be uniformly charged. The patterns that emerge spontaneously in certain kinds of chemical reactions governed by the diffusion of the individual components provide an equally rewarding visual experience, especially the acute sensitivity to variations in the rate parameters governing the process.

f) Chaos is yet another subject that owes much of its existence to computer graphics. Here again the role of visual exploration is a an essential one. One of the classical examples is the Lorenz attractor, and it is important to be able to probe this problem – with its characteristic butterfly-shaped trajectories – in all its 3D glory.

g) The final topic that will be mentioned here is that of cellular automata. These simple models can be used in studying a wide variety of problems, including growth phenomena, models with a more biological flavor (such as the game of Life), as well as simulations of fluid dynamics. Here the visualization can focus on the patterns formed by the states of the individual cells, or in the case of fluids, on the nature of the flow field itself.

3. Design considerations

The problem in designing computer software capable of demonstrating phenomena such as those enumerated above lies not so much in programming the computations associated with the mathematical and algorithmic details of the problem but rather with providing the required graphical and interactive capabilities. In early implementations of these demonstrations using low-power computers the design process also entailed minimizing the computational effort, such as by avoiding floating-point arithmetic, but such extremes are no longer necessary.

While the computations can be programmed in some standardized language, such as C, and are fully portable between different computer systems, the programming required to provide the graphics and interactivity is very much system dependent. Fortunately, standards (not necessarily ideal ones) are emerging, and in the case of Unix-based systems the availability of the X window environment for graphics and Motif for interface development help ease this problem. For more advanced graphics, GL and OpenGL software is available on some platforms, and these are designed to make use of certain specialized graphics hardware, if installed. But while the underlying logic of the software needed for the visualization and user interfacing remains the same irrespective of the platform, the actual implementation on a machine such as the Macintosh is very different from X/Motif. Given enough experience it is possible to construct function libraries that hide much, if not all of the machine-dependent detail, but this is no easy task and any new application is likely to demand extensions to whatever functionality has already been included in such a library. (There are commercially available software tools to aid the graphical user interface design process, but the author has not had any experience with them.)

A well-designed interface should be simple and intuitive, and if a set of programs is being developed, also consistent. The user should be aware of what operations are allowed at any time and the type and range of data that must be supplied to the program. All data must be checked for validity before being accepted – in fact it is preferable that the user not actually enter numerical values by typing but rather by selecting from a menu of allowed choices or by using some graphical 'analog' device, such as slider or thumbwheel, whose setting can be changed by a simple mouse dragging operation. Other control over what the program is doing can employ menu selections, use of graphical pushbuttons, and direct manipulation of on-screen graphics using the mouse (such as in establishing the initial state for a cellular automata study or in changing the viewpoint for 3D visualization). Some form of online documentation integrated into the program – possibly context dependent – is a definite benefit.

In order for the user to appreciate the interactive nature of the program it is vital that prompt responses appear following any user input. This means that a lengthy computation must be fragmented, with user input being accepted and processed frequently (preferably several times a second) throughout the course of the calculation. If, for some reason, the program is temporarily unable to accept input, some visual cue (such as a changed cursor shape) should be provided.

It is important that if animated graphics are used the changes to the imagery should occur gradually and without any display flashing. This requires tuning both the rate at which the computation progresses and the rate at which the display is updated. To provide a smooth transition between successive images and avoid annoying flashes it is generally necessary to resort to double buffering, where the image is created out of sight and then rapidly transferred to the screen; how this is achieved is very much system dependent.

4. Summary

In this all too brief article we have attempted to convey a feeling of what interactive visualization can offer the statistical physicist, as well as some idea of the effort involved in actually developing such software. There can be little doubt as to the pedagogical importance of this kind of approach, and increasing computer power (both for computation and graphics) will serve to increase its role in the future. The provision of graphic output and interactivity does increase the software development effort, but once available, such features prove to be of considerable value.

Bibliography

Earlier articles by the author describing software of this kind for the Amiga and Macintosh computers are as follows:

D.C. Rapaport, Computers in Physics **3**, 5:18 (1989).

D.C. Rapaport, J. Stat. Phys. **58**, 775 (1990).

D.C. Rapaport and M. Meyer, in *Fractals in Science*, eds. A. Bunde and S. Havlin (Springer, 1994), p.257.

Further information on simulation, visualization, and other topics mentioned in this article can be found in the following books:

M.P. Allen and D.J. Tildesley, *Computer Simulation of Liquids* (Oxford, 1987).

A. Bunde and S. Havlin, eds., *Fractals in Science* (Springer, 1994).

P. Cvitanovic, ed., *Universality in Chaos* (Adam Hilger, 1984).

S. Garfinkel, et al, *The Unix-haters Handbook* (IDG Books, 1994).

H. Gould and J. Tobochnik, *An Introduction to Computer Simulation Methods* (Addison-Wesley 1988).

H-O. Peitgen and D. Saupe, eds., *The Science of Fractal Images* (Springer, 1988).

D.C. Rapaport, *The Art of Molecular Dynamics Simulation* (Cambridge, 1995).

R.S. Wolff and L. Yaeger, *Visualization of Natural Phenomena* (Springer, 1993).

S. Wolfram, *Theory and Applications of Cellular Automata* (World Scientific, 1986).

First Versus Second Order Phase Transitions in the Three-Dimensional Three-State Potts Model in Random Fields

K. Eichhorn and K. Binder

Institut für Physik, Johannes Gutenberg Universität Mainz,
Staudinger Weg 7, D-55099 Mainz, Germany

Abstract. The ordering of three-states Potts ferromagnets on the simple cubic lattice exposed to random fields is investigated by extensive Monte Carlo simulations. Evidence is presented that the transition is second order for intermediate strength of the fields, while it presumably is first order for large field strength. The implications for various theoretical predictions are briefly discussed.

1. Introduction

The Potts model [1] deals with a Hamiltonian, where each lattice site carries a spin s_i that may take any out of q posssible states, and an energy J is gained if two neighboring sites are in the same state. Of particular interest is that the order of the phase transition depends on both q and the dimensionality d of the system: the transition is first order when the system lies above a separatrix $q_c(d)$ in the (q, d)-plane while it is second order below (Fig. 1).

We consider an extension of the model where a quenched field is added that randomly favors the state h_i ($h_i = 1, \ldots, q$),

$$\mathcal{H}_{\mathrm{RFPM}} = -J \sum_{<i,j>} \delta_{s_i s_j} - H_{\mathrm{RF}} \sum_i \delta_{s_i h_i} \, . \tag{1}$$

Here, H_{RF} denotes the strength of the random field and $< i, j >$ runs over pairs of neighboring spins. Now, Blankschtein et al. [2] made very interesting predictions on the critical behavior of that model, applying a (modified) principle of "dimensional reduction" [3-5]: the critical behavior of the random field system in d_{RF} dimensions should be identical to that of a pure system in d dimensions, with $d_{\mathrm{RF}} = d + 2 - \eta(d)$. Consequently, the separatrix in the (q, d)-plane gets shifted (broken curve in Fig. 1). The lower critical dimension (where $q_c(d \to \mathrm{lcd}) \to \infty$) is raised from $d = 1$ to $d = 2$, while the upper critical dimension (where $q_c(d \to \mathrm{ucd}) \to 2$: remember that $q = 2$ means an Ising model!) gets shifted from $d = 4$ to $d = 6$. The physically most interesting case, $q = 3$ in $d = 3$, which was above the separatrix in the pure case (a weak first order transition is in fact observed then, see e.g. [6]), now falls well below the separatrix: for the RFPM {Eq. (1)} second order transitions should be possible. A phase diagram scenario has been proposed [7] where the phase diagram in the (H_{RF}, T) plane for $q_c(d) < q < q_c^{\mathrm{RF}}(d)$ exhibits two tricritical

Springer Proceedings in Physics, Vol. 80
Computer Simulation Studies in Condensed-Matter Physics VIII
Editors: D.P. Landau · K.K. Mon · H.-B. Schüttler

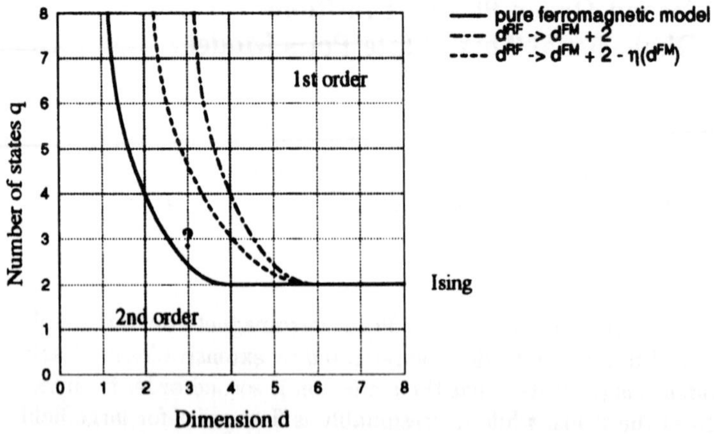

Figure 1: Schematic plot of $q_c(d)$. After Blankschtein et al. [2].

points. The tricritical point at higher temperature starts out at $q = q_c(d)$ for $H_{RF} = 0$, while at $q = q_c^{RF}(d)$ it merges with the low temperature triple point (which also exists in the Ising case, $q = 2$ [8]), and thus for $q > q_c^{RF}(d)$ only first order transitions remain. However, series expansion analyses [9] have claimed that for $q = 3, d = 3$ the transition stays first order throughout, implying that Fig. 1 is meaningless. Clarification of this longstanding controversy is a goal of our work.

2. Simulation and Analysis Methods

We apply multispin coding techniques [10] and histogram reweighting [11]. On a Cray YMP processor the program performs $21.2 \cdot 10^6$ spin updates per second, on an Intel PARAGON I860, where the "quenched averaging" can be easily done in parallel, the speed reaches $1.3 \cdot 10^6$ per node. Typical times for equilibration used are 10^5 to 10^6 Monte Carlo steps (MCS) per spin. For a medium strength of the random field ($H_{RF} = 1$) sizes up to $L = 28$ can be equilibrated at and above T_c, and hence a meaningful finite size scaling analysis [12] is possible (Sec. 3). For strong random fields, such as $H_{RF} = 2$, however, it is no longer possible to equilibrate systems as small as $L = 10$ at T_c [7], and large systems reveal pronounced hysteresis. Cooling below the transition the system cannot order in a monodomain sample at the given observation times, rather an irregular structure of large domains is frozen in. We have found it useful to analyse the distribution of the magnetization in subblocks of linear dimension l_{box}, varying l_{box} from 4 to 20 for $L = 50$ (Fig. 2). For $H_{RF} = 1$ one observes two distinct peaks, as expected for first order transitions [6]. The lower peak results from the fact that the order parameter is two-component and moves to zero as $l_{box} \to \infty$; this peak represents the disordered phase. The upper peak

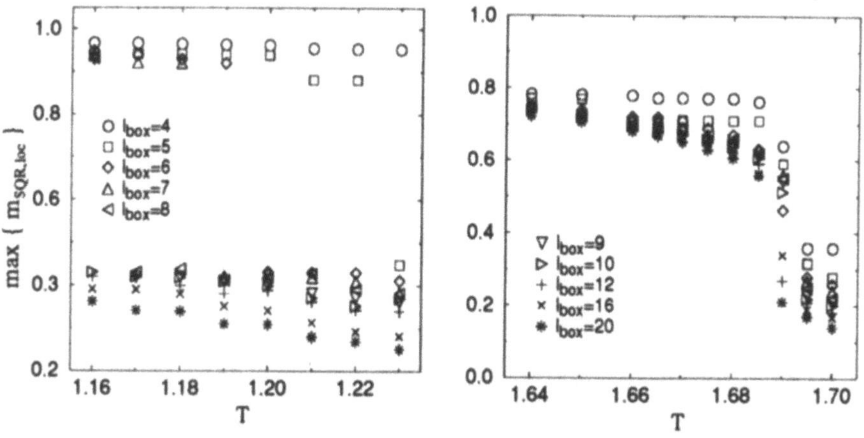

Figure 2: Plot of the distribution maxima of the absolute value of the magnetization versus temperature (in units of the exchange energy in simplex representation) for $H_{\mathrm{RF}} = 2$ (left side) and $H_{\mathrm{RF}} = 1$ (right side) averaged over 9 random field configurations

stabilizes with increasing l_{box} for $T < 1.2$ at a large value ≥ 0.9. Note that the ordered peak is only resolved clearly in rather small boxes - otherwise they contain domain walls in most cases, and a broad smeared-out distribution results. The corresponding distribution for the medium field strength $H_{\mathrm{RF}} = 2$ shows a totally different behavior: no double peak structure can be found indicating a continuous transition.

3. Finite Size Scaling Analysis for $H_{\mathrm{RF}} = 1$

Both energy and magnetization have a very smooth dependence on temperature and this does not sharpen up much with increasing lattice size. This indication of a second order transition is corroborated by the specific heat C and susceptibility χ [7]: both quantities show peaks which grow rather slowly. A log-log plot of peak heights versus lattice size clearly exhibits two regimes. For $L < 12$ one observes effective exponents that have similar values as the pure Potts ferromagnet for such small sizes [6]. But in the second regime (for $L > 12$) now clearly smaller slopes are observed, different from the pure case: in the latter a crossover to the large slope corresponding to the exponent $d = 3$ is observed, as it should be for a first order transition [6,12].

To find T_c we have extrapolated various characteristic temperatures $T_c(L)$, { i.e. location of maxima of C and χ, minima and zero of the forth-order cumulant g_L, etc. [7] } versus $L^{1/\nu_{\mathrm{eff}}}$ adjusting ν_{eff} so that one gets an optimal fit to a straight line. This yields $T_c = 1.6905 \pm 0.010$ and $\nu_{\mathrm{eff}} \approx 0.65$. This result of T_c is also compatible with the cumulant intersections (Fig. 3). One sees there

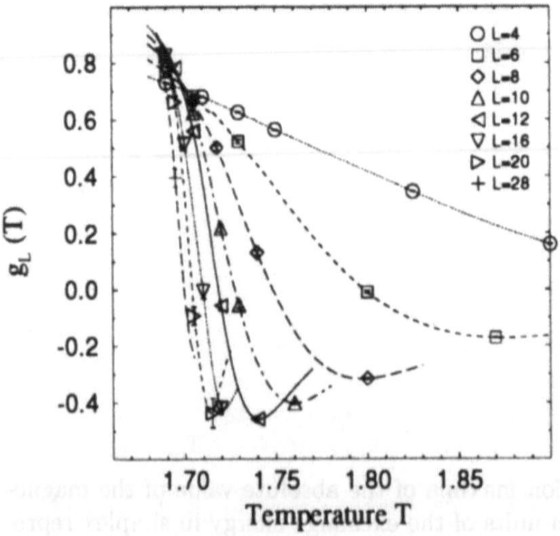

Figure 3: Cumulant $g_L(T)$ plottet versus temperature for $H_{RF} = 1$ and various system sizes as indicated.

that $g_L(T)$ {which is normalized such that $g_L(T \to \infty) = 0$, $g_L(T = 0) = 1$ [6]} has a minimum at negative values above T_c. But while this minimum in the pure case [6] gets deeper and deeper as L increases, here the trend changes for $L \approx 12$, and for larger L the minimum becomes shallower again. The intersection values also shift from $g_L^* \approx 0.7$ at small sizes (compatible with the pure case [6]) to rather high values ($g_L^* \approx 0.9$ for $L = 16$ to 20) and an even larger value is conceivable for larger L.

Knowing T_c one can estimate the finite size behavior of order parameter $[<|m|>]_{av}$ and disconnected susceptibility $k_B T \chi_{dis} \equiv L^d \left[<m>^2\right]_{av}$ at T_c (Fig. 4). We find $[<|m|>]_{av} \sim L^{-\beta/\nu}$ with $\beta/\nu = 0.34 \pm 0.01$ (note that this error does not include systematic errors due to a possible misjudgement of T_c). Fig. 4 shows that for $L > 12$ χ and χ_{dis} exhibit different exponents $\bar{\gamma}/\nu = 2.49 \pm 0.38$ and $\gamma/\nu = 1.42 \pm 0.12$. Thus the hyperscaling law $d = \gamma/\nu + 2\beta/\nu$ indeed seems violated, while a related law [7] $d = \bar{\gamma}/\nu + 2\beta/\nu$ holds within error. Note that this hyperscaling violation has subtle consequences for the probability distribution $P_L(m, T)$ [7]. In a finite size scaling context, the modified hyperscaling law [4,5] $\gamma + 2\beta = \nu(d - 2 + \eta)$ can be understood most simply. First one notes the exact inequality $\left[<m^2>\right]_{av} \geq \left[<m>^2\right]_{av} \geq [<m>]_{av}^2$ and assumes as usual, that for the critical exponents of these terms the inequality is an equality. Since at T_c we have $[<m>]_{av} \sim L^{-\beta/\nu}$, it follows that the disconnected susceptibility $\chi_{dis} \sim L^d \left[<m>^2\right]_{av} \sim L^{d-2\beta/\nu} = L^{\bar{\gamma}/\nu}$ with $\bar{\gamma}/\nu + 2\beta/\nu = d$ the standard hyperscaling relation. Second we state that in a volume L^d the excess magnetic field energy of one state is of order $H_{RF} L^{d/2}$, and hence a magnetization per spin $<m> = \chi H_{RF} L^{-d/2}$ results.

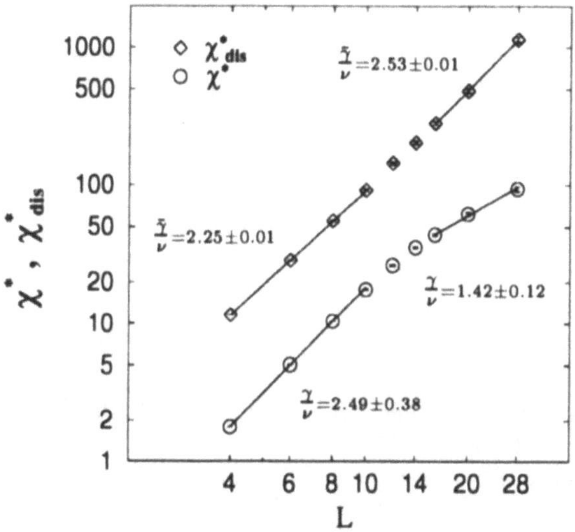

Figure 4: Log-log plot of the maxima χ^* of $k_B T\chi \equiv L^d\left[< m^2 > - < |m| >^2\right]_{av}$ (lower set of data) and of the disconnected susceptibility $\chi^*_{\rm dis}$ (upper set of data)

Using $< m >\sim L^{-\beta/\nu}$, $\chi \sim L^{\gamma/\nu}$ at T_c yields $d/2 = \gamma/\nu + \beta/\nu$ and hence $\bar{\gamma} = 2\gamma$. With $\gamma/\nu = 2 - \eta$ one retrieves the modified hyperscaling law [4,5] $\gamma + 2\beta = \nu(d - 2 + \eta)$.

4. Discussion

In summary, evidence has been presented that the RFPM has a first order transition for strong random fields, but a second order transition for intermediate field strength, compatible with the suggestions of Blankschtein et al. [2]. The critical exponents of the RFPM do indeed seem to violate hyperscaling, but presumably hyperscaling holds with the exponent $\bar{\gamma}$ of the disconnected susceptibility. Our data are probably not in the asymptotic regime, and hence cannot conclusively test the relation $\bar{\gamma} = 2\gamma$ [4,5], however. Also a precise location of the proposed two tricritical points [7] does not yet seem feasible with present computer resources. A finite size scaling interpretation of the relation $\bar{\gamma} = 2\gamma$ has been given.

Acknowledgements

One of us (K. E.) is supported by the Deutsche Forschungsgemeinschaft (DFG), Sonderforschungsbereich 262/D1. We are grateful to the Höchstleistungsrechenzentrum Jülich (HLRZ) and to the Regionales Hochschulrechenzentrum Kaiserslautern (RHRK) for generous grants of computing time. K. B. thanks A. Aharony, M. Schwartz and J. Adler for stimulating discussions. We are also indebted to J.D. Reger for a helpful cooperation in the early steps of this work.

References

[1] F.Y. Wu, *Rev. Mod. Phys.* **54**, 235 (1982)
[2] D. Blankschtein, Y. Shapir and A. Aharony, *Phys. Rev. B* **29**, 1263 (1984)
[3] A. Aharony, Y. Imry and S.K. Ma, *Phys. Rev. Lett.* **37**, 944 (1976)
[4] M. Schwartz, *J. Phys. C* **85**, 135 (1985);
M. Schwartz und A. Soffer, *Phys. Rev. Lett.* **55**, 2499 (1985)
[5] M. Schwartz, M. Gofman and T. Nattermann, *Physica A* **178**, 6 (1991)
[6] K. Vollmayr, J. Reger, M. Scheucher and K. Binder, *Z. Phys. B* **91**, 113 (1993)
[7] K. Eichhorn and K. Binder, *preprints*
[8] A. Aharony, *Phys. Rev. B* **18**, 3328 (1978)
[9] Y.Y. Goldschmidt and G. Xu, *Phys. Rev. B* **32**, 1876 (1985);
Nucl. Phys. B **265**, 1 (1986)
[10] G. Bhanot, D. Duke and R. Salvador, *J. Stat. Phys.* **44**, 985 (1986)
[11] A.M. Ferrenberg and R.H. Swendsen, *Phys. Rev. Lett.* **61**, 2635 (1988); ibid. **63**, 1195 (1989)
[12] K. Binder, in *Computational Methods in Field Theory* (H. Gausterer and C.B. Lang, eds.) p. 59 (Springer, Berlin 1992), and references therein.

Detailed Balance Method of Chemical Potential Determination

P.J. Fay[1], *J.R. Ray*[2] *and R.J. Wolf*[3]

[1]Office of the Vice-Provost of Research, University of South Carolina, Columbia, SC 29208, USA
[2]Department of Physics and Astronomy, Kinard Laboratory of Physics, Clemson University, Clemson, SC 29634-1911, USA
[3]Savannah River Technology Center, Westinghouse Savannah River Laboratory, Aiken, SC 2908, USA

Abstract. We present a method of determining the chemical potential in Monte Carlo or molecular dynamics simulations by determining the conditions for detailed balance between fictitious creation and destruction attempts.

1. Introduction

We have developed a new way of determining the chemical potential in Monte Carlo or molecular dynamics simulations which is nondestructive and makes use of detailed balance between fictitious creation and destruction attempts. The method has been shown to work at densities that are at least 10% higher than other methods that have been presented. The method makes use of a pairing between ensembles which relates each closed ensemble with a detailed balance method (DBM) associate. The best known pairing is the canonical ensemble, TVN ensemble, with the grand canonical ensemble, $TV\mu$ ensemble, here T is the temperature, V the volume, N the particle number and μ chemical potential of the system. The other DBM pairs for a single component system are the EVN ensemble with the $LV\mu$ ensemble the HPN ensemble with the $RP\mu$ ensemble, and the TPN ensemble with the $TP\mu$ ensemble. The closed particle simulation, e.g. the one with N fixed, can be carried out using either the molecular dynamics or Monte Carlo method. These pairings can be most easily determined from the Graben-Ray ensemble classification scheme [1] and are all presented in [2]. Since we use the Monte Carlo method for the creation/destruction acceptance probabilities in all cases the acceptance probability for a creation is $W(x_N \rightarrow x_{N+1}) = \min[1, P(x_{N+1})/P(x_N)]$ where $P(x_N)$ is the probability of finding N particles at the positions x_N. Likewise for the destruction probability $W(x_N \rightarrow x_{N-1}) = \min[1, P(x_{N-1})/P(x_N)]$. As an example for the HPN -$RP\mu$ pair we have the$RP\mu$ creation results

$$\frac{P(x_{N+1})}{P(x_N)} = \frac{bV\Gamma(3N/2)}{(N+1)\Gamma(3(N+1)/2)} \frac{\{R - P_{ext}V - U(x_{N+1}) + \mu(N+1)\}^{3(N+1)/2-1}}{\{R - P_{ext}V - U(x_N) + \mu N\}^{3N/2-1}}, \quad (1)$$

Springer Proceedings in Physics, Vol. 80
Computer Simulation Studies in Condensed-Matter Physics VIII
Editors: D.P. Landau · K.K. Mon · H.-B. Schüttler

where $b=(2\pi m)^{3/2}/h^3$, with m the particle mass, h Planck's constant, Γ the Gamma function, R the energy, P_{ext} the external pressure, and U the potential energy. For the destruction attempts one has a similar formula with $N+1\rightarrow N-1$ in the numerator of eq. (1). The DBM method involves carrying out creation/destruction attempts over a chemical potential grid using the open ensemble probabilities and keeping track of the success of these attempts. A plot of the number of successful creation/destruction attempts versus the chemical potential gives creation and destruction curves. The intersection of these curves give an estimate of the chemical potential of the system. It is important that the creation and destruction attempts are not applied to the system, that is these are fictitious and the DBM is nondestructive. This makes the DBM a very efficient method since the creation/destruction energies can be calculated one time each chemical potential sweep and then used for all chemical potentials in the grid, we used typically 1000 chemical potentials in the grid.

2. Results and Discussion

When one starts the DBM process one does know the value of the chemical potential, therefore, a chemical potential sweep over a large chemical potential grid is carried out first. In fig. 1 we show the creation/destruction curves for a Lennard-Jones system at a reduced temperature of 1.2 and a reduced density of 0.85. The creation curve is on the extreme right of this graph and just barely shows up. We can see from these curves that the intersection is near -0.1 eV. Narrowing down the range of chemical potential variation we collect additional DBM statistics and obtain fig. 2 which shows the intersection of the creation and destruction curves for this case. By fitting smooth curves to the creation/destruction curves we obtain a chemical potential of -0.09575 eV for this case.

One important property of the DBM is that we can carry out a verifying simulation by using the value determined for the chemical potential in a open ensemble simulation to see if N changes from the closed ensemble value. Carrying out this procedure using the grand canonical ensemble, for the system discussed in figs. 1 and 2, we find for a 500 particle system that $\langle N \rangle = 497.86 \pm 1.52$, where the error estimate has been obtained by carrying out 5 Monte Carlo simulations of 20 thousand moves each. Results in the literature are consistent with this value for the chemical potential for the Lennard-Jones system at the given state conditions. Further detailed results may be found in ref. 2.

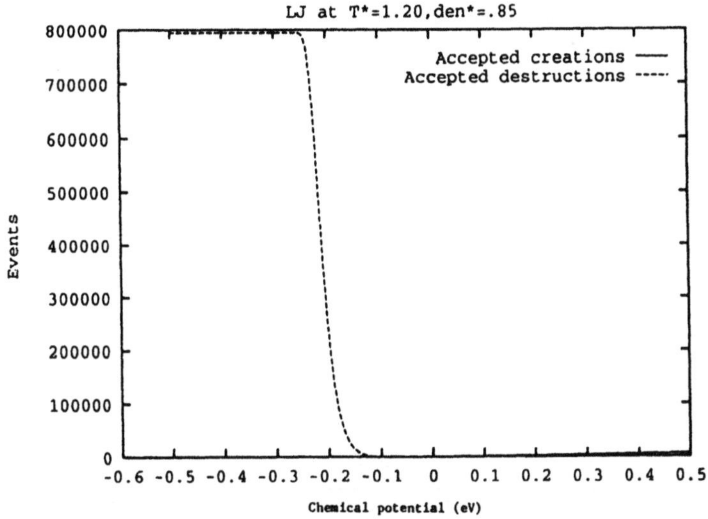

Figure 1. The creation destruction curves for the Lennard-Jones system for a reduced temperature of 1.2 and a reduced density of 0.85. Plotted is the total number of accepted creation/destruction events. The creation curve is just barely visible at the bottom right hand side of the figure.

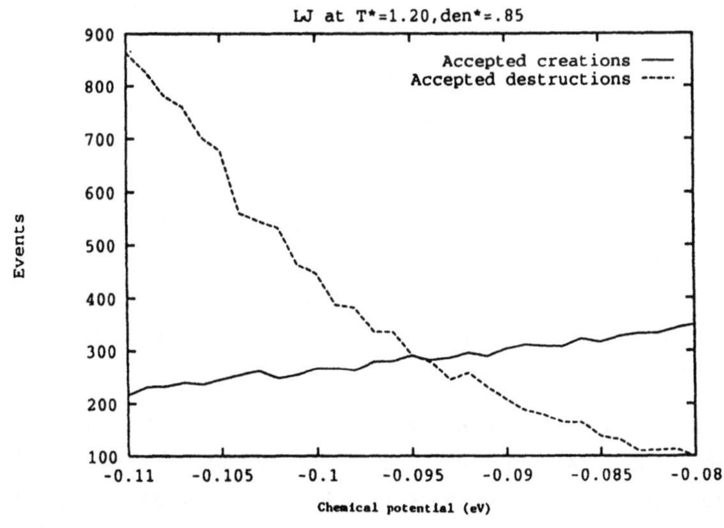

Figure 2. The creation destruction curves for the same system as in fig. 1 plotted on a finer chemical potential grid.

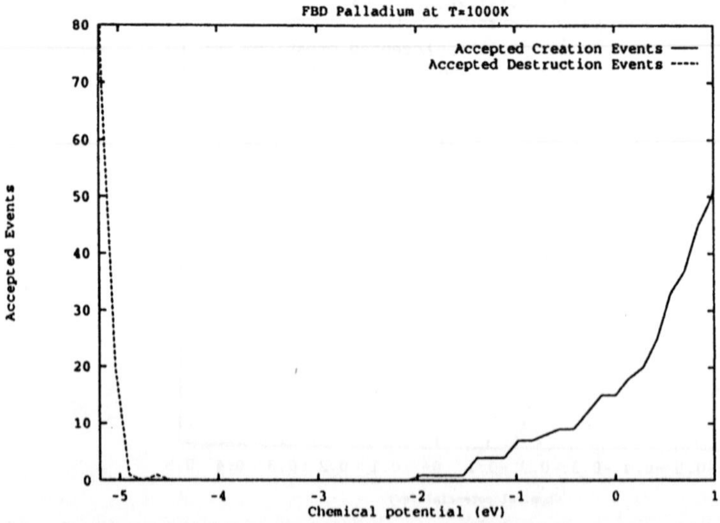

Figure 3. The creation/destruction curves for FCC palladium illustrating the fact that the DBM method will not work at low enough temperature and/or high enough density.

We were able to determine the chemical potential of the Lennard-Jones system along the 1.2 reduced temperature isotherm for reduced densities of at least 1.0 which is 10% higher than any density we found results for in the literature. This suggests that the DBM is more efficient than other methods for determining the chemical potential, such as for example, Widom's test particle method. We feel the reason the DBM is a more efficient method (meaning it works at higher densities and lower temperatures) is that when we obtain a successful creation or destruction it goes in the creation/destruction curves with the same weight even if it is an improbable event; in e. g. Widom's method an improbable event gives no contribution to the chemical potential since the energy change appears in a negative exponent.

As a final point we mention that at low enough temperatures or at high enough densities the DBM will fail because the statistics for successful creation/destruction events is too poor to allow an intersection of the creation/destruction curves. In fig. 3 we show such a situation for solid palladium modeled by an embedded atom model potential. This is a FCC solid at 1000 K. Refining the chemical potential grid does not lead to an intersection of the creation/destruction curves in this case. We have developed other methods that are successful in such cases.

Acknowledgment

We acknowledge support from the U.S. Department of Energy under contract No. DE-AC09-88SR18035 and through SCUREF.

References

1. J. R. Ray and H. W. Graben, Mol. Phys. 80, 1183 (1993).
2. P. J. Fay, J. R. Ray, and R. J. Wolf, J. Chem. Phys. 100, 2154 (1994).

Monte-Carlo Simulation of the Casimir Effect in Critical Strips

M. Krech and D.P. Landau

Center for Simulational Physics, The University of Georgia, Athens, GA 30602, USA

If a critical system is confined to a finite geometry, critical fluctuations of the order parameter generate long-ranged forces between the system boundaries. These forces, commonly known as Casimir forces, are characterized by universal amplitudes and scaling functions. A hybrid Monte-Carlo algorithm has been devised and used to measure the Casimir amplitudes directly and accurately. We apply the algorithm to a critical q-state Potts model confined to a rectangular $L_x \times L_y$ geometry for various boundary conditions and aspect ratios $s = L_y/L_x$ and find good agreement with rigorous results.

1. Introduction

One of the most fundamental discoveries in the theory of electomagnetism dates back to the year 1948, when the Dutch physicist H.B.G. Casimir realized that two metallic plates placed parallel to one another at a distance L in vacuum experience an attractive force [1]. According to the modern view of this effect, which had also been proposed by Casimir, the metallic plates impose boundary conditions on the *zero-point* fluctuations of the electromagnetic field thereby imposing an L dependence on the fluctuation spectrum. This in turn gives rise to an L dependence in the total *energy* E of the configuration. Likewise, *thermal* fluctuations of the electromagnetic field at high temperatures introduce a similar L dependence to the configuration energy [2].

Several years after Casimirs discovery a general theory of these so-called dispersion forces in layered dielectrics was formulated [3] yielding a unified picture of the Casimir forces and the well known van-der-Waals forces [4]. It was already noted in Ref.[3] that dispersion forces play an important role in the vast realm of wetting phenomena [5]. In the simplest case a wetting transition leads to the formation of a macroscopically thick liquid layer on the surface of a homogeneous plane substrate which is exposed to the vapor of the wetting agent. The combined substrate-liquid-vapor system then has the structure of layered dielectrics considered in Ref.[3]. The equilibrium thickness of the wetting layer is determined by the minimum of the so-called effective interface potential Ω as function of a *prescribed* thickness l [5]

$$\lim_{A\to\infty} \frac{\Omega(l)}{A} = \omega(l) = l\,(\rho_l/\rho_v - 1)\, p_0(T)\, \delta p + \sigma_{sl} + \sigma_{lv} + \delta\omega(l), \tag{1.1}$$

where A is the interfacial area, ρ_l and ρ_v denote the liquid and the vapor density, respectively, $p_0(T)$ is the vapor pressure *at* liquid-vapor coexistence, and δp is a dimensionless measure of the undersaturation of the vapor. The substrate-liquid and liquid-vapor interfacial tensions σ_{sl} and σ_{lv} do not depend on l and $\delta\omega(l)$ contains the contribution of the van-der-Waals forces which has the asymptotic form

$$\delta\omega_{vdW}(l) = \begin{cases} W\, l^{-2} + O(l^{-3}) & \text{(nonretarded)} \\ W_r\, l^{-3} + O(l^{-4}) & \text{(retarded).} \end{cases} \tag{1.2}$$

According to the mechanism underlying the Casimir effect in electromagnetism critical fluctuations of the order parameter *at* a critical point of a wetting agent generate an additional long-ranged force which contributes to $\delta\omega(l)$ (see Eq.(1.1)). One finds [6]

Springer Proceedings in Physics, Vol. 80
Computer Simulation Studies in Condensed-Matter Physics VIII
Editors: D.P. Landau · K.K. Mon · H.-B. Schüttler

$$\delta\omega(l) = \delta\omega_{vdW}(l) + k_B T_c \Delta l^{-(d-1)} \tag{1.3}$$

in d dimensions, where Δ is the *universal* Casimir amplitude and T_c is the critical temperature. Therefore one expects a critical effect on the equilibrium layer thickness L, e.g., in the vicinity of the λ-transition in ^{4}He [7] and near critical end points of the demixing transition in binary mixtures [8, 9]. The Casimir amplitude Δ in Eq.(1.3) is exactly known for the critical Ising and the 3-state Potts model in $d = 2$ dimensions [10]. However, in $d = 3$, which is the experimentally relevant case, Δ can only be estimated [6, 11].

2. The Monte–Carlo method

In order to be able to compare our Monte–Carlo estimates for Δ with as many exact results as possible we consider the well-known ferromagnetic q–state Potts model defined by the Hamiltonian

$$\mathcal{H}\{\sigma\} = -J \sum_{\langle \mathbf{r},\mathbf{r}'\rangle} \delta_{\sigma(\mathbf{r}),\sigma(\mathbf{r}')} \tag{2.1}$$

on a square lattice with the usual nearest neighbor Kronecker-δ interactions confined to a rectangular geometry with $L_x \times L_y$ lattice sites and $L_x \geq L_y$. Periodic boundary conditions are applied along the x–direction and various other boundary conditions are applied along the y–direction. Following Ref.[12] we probe the L_y dependence of the free energy of the Potts model for even L_y by introducing a seam Hamilonian $\mathcal{H}_s\{\sigma\}$ between the two rows of spins at $y = \frac{1}{2}L_y$ and $y = \frac{1}{2}L_y + 1$ such that the *combined* Hamiltonian $\mathcal{H}\{\sigma\} + \mathcal{H}_s\{\sigma\}$ describes two uncoupled Potts models each on a square lattice with $L_x \times \frac{1}{2}L_y$ sites. One of these subsystems is endowed with periodic boundary conditions along the y–direction whereas the other inherits the boundary conditions of the initial system described by Eq.(2.1). If, for example, periodic boundary conditions along the y–direction are incorporated in $\mathcal{H}\{\sigma\}$ then $\mathcal{H}_s\{\sigma\}$ has the form

$$\mathcal{H}_s\{\sigma\} = J \sum_{x=1}^{L_x} \left[\delta_{\sigma(x,L_y/2),\sigma(x,L_y/2+1)} + \delta_{\sigma(x,1),\sigma(x,L_y)} - \delta_{\sigma(x,1),\sigma(x,L_y/2)} - \delta_{\sigma(x,L_y/2+1),\sigma(x,L_y)}\right]. \tag{2.2}$$

The corresponding change in the free energy $\mathcal{F}$ at $T = T_c$ is simply given by

$$\mathcal{F}\{\mathcal{H} + \mathcal{H}_s\} - \mathcal{F}\{\mathcal{H}\} = -k_B T_c \langle \exp(-\mathcal{H}_s / k_B T_c)\rangle_{\mathcal{H}}, \tag{2.3}$$

where $\langle \ldots \rangle_{\mathcal{H}}$ denotes a thermal average with respect to the Hamiltonian $\mathcal{H}$ of the initial Potts model. In the limit of vanishing aspect ratio $s = L_y/L_x$ either of the free energies in Eq.(2.3) can be written in the form of Eq.(1.1) with Ω replaced by $\mathcal{F}$ and A replaced by L_x. According to the choice of the boundary conditions for the split system $\mathcal{F}\{\mathcal{H}\}$ and $\mathcal{F}\{\mathcal{H} + \mathcal{H}_s\}$ have identical bulk and surface contributions which therefore cancel each other in Eq.(2.3). Moreover, logarithmic corrections to critical interfacial tensions due to *finite* L_x [13] cancel as well. By evaluating the r.h.s. of Eq.(2.3) we thus probe the critical finite–size part of the free energy *directly*. Note, that due to the extreme short ranged nature of the interaction in the Potts model we have $\delta\omega(L_y) = k_B T_c \Delta L_y^{-1}$ (see Eq.(1.3)) apart from *exponentially* small terms.

In order to obtain an accurate measurement of the r.h.s. of Eq.(2.3) special sampling techniques are required [12]. We therefore consider a sequence of Hamiltonians $\mathcal{H}_a \equiv \mathcal{H} + a\mathcal{H}_s$ for typically 9–12 values of a ($0 \leq a \leq 1$) in our simulation. For each a $\mathcal{H}_a$ describes a ferromagnetic Potts model with additional bonds (see Eq.(2.2)) which can be updated using the standard Metropolis or Wolff algorithm. In our simulation we use a hybrid scheme (with the R1279 random number generator [14]) which consists of one Metropolis update of the whole lattice and a subsequent Wolff update in order to keep the autocorrelation times from growing too quickly with the system size. The values for a are chosen such that the histograms $H_a(E_s)$ of the seam energy $E_s = \mathcal{H}_s\{\sigma\}$ for successive values of a have a sufficiently large overlap. Note, that according to Eq.(2.2) E_s grows with L_x so that more histograms are needed for larger L_x. For the data analysis we apply the optimized multiple histogram analysis [15] which yields the normalization coefficient $z(a)$ for each histogram

$H_a(E_s)$, where $z(1)/z(0) = \langle \exp(-\mathcal{H}_s/k_BT_c)\rangle_{\mathcal{H}}$ is the desired quantity. The statistical error is determined from the variance of the resulting density of states $W(E_s)$ [15]. The autocorrelation times needed for the analysis have been estimated from the time diplaced autocorrelation function $\langle E_s(t)E_s(0)\rangle_{\mathcal{H}_a}$ as functions of a for $q = 2, 3, 4$.

3. Results and discussion

Our simulation are performed on *finite* lattices and therefore a sufficiently large value for L_y has to be found which enables us to neglect lattice effects in the finite–size part of the free energy, and a sufficiently large value for L_x must be identified which brings our rectangular geometry sufficiently close to the infinite strip limit. In order to accomplish both we first concentrate on the Ising limit $q = 2$ of the Potts model for purely periodic boundary conditions which has been rigorously solved on finite lattices [16]. Defining a Casimir amplitude $\Delta_{per}(s)$ for periodic boundary conditions as a function of s by $\Delta_{per}(s) = s/3 \ln\langle \exp(-\mathcal{H}_s/k_BT_c)\rangle_{\mathcal{H}}$ we use of Eq.(3.37) in Ref.[16] to evaluate $\Delta_{per}(s)$ in the continuum limit as a function of s. The direct comparison between this exact result and our Monte–Carlo simulation with 6.48×10^6 hybrid Monte–Carlo steps for each of the system sizes $40 \leq L_x \leq 320$ for $L_y = 40$ is shown in Fig.1. The excellent agreement leads us to the conclusion that $L_y = 40$ is sufficiently close to the continuum limit and that $s = 1/8$ is sufficiently close to the infinite strip limit $s = 0$ within our error bars.

In order to study the decay of lattice effects with increasing L_y we evaluate $\Delta_{per}(s = 1/8)$ for a sequence of lattices with $12 \leq L_y \leq 40$ again using 6.48×10^6 hybrid Monte–Carlo steps each. For a critical q–state Potts model in an infinite strip the exact Casimir amplitude is $\Delta_{per} = -\frac{\pi}{6}c$, where c is the conformal anomaly number ($c = \frac{1}{2}$ for $q = 2$ and $c = \frac{4}{5}$ for $q = 3$ [10]). The 4–state Potts model is a borderline case, which is believed to be characterized by $c = 1$ [17]. The Casimir amplitude Δ_{per} for $q = 4$ should therefore be the same as for a critical single–component Gaussian model in $d = 2$. The comparison with the above expectation is shown in Fig.2. Within the error bars the estimated values for Δ_{per} approach the theoretically expected values monotonically from below as L_y increases. Using the above relation between Δ_{per} and c we estimate $c = 1.006 \pm 0.008$ for $q = 4$ from the last four datapoints which is in very good agreement with the expected value.

Preliminary simulations using antiperiodic boundary conditions for $q = 2$ and their analogue for $q = 3$ also show very good agreement with exact results for the Casimir amplitudes (see Ref.[10]).

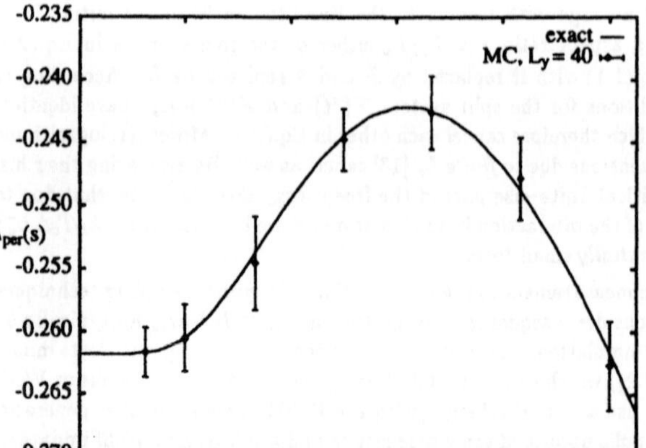

Fig.1: Casimir amplitude $\Delta_{per}(s)$ (see main text) as a function of the aspect ratio s for $q = 2$ and periodic boundary conditions. The solid line shows the exact result in the continuum limit (see Ref.[16]) and the points indicate Monte–Carlo results for $s = \frac{1}{8}, \frac{1}{5}, \frac{1}{3}, \frac{1}{2}, \frac{2}{3}, \frac{5}{6}$, and 1.

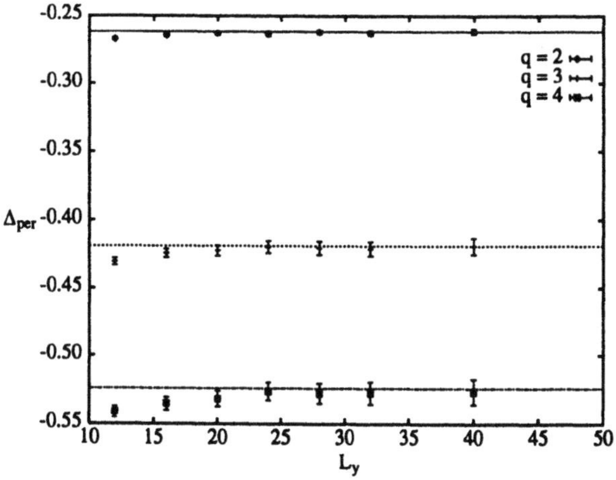

Fig.2: Casimir amplitude Δ_{per} for fixed aspect ratio $s = \frac{1}{8}$ as a function of L_y. The theoretical values are indicated by the solid line ($q = 2$), the dotted line ($q = 3$), and the dash–dotted line ($q = 4$).

The statistical errors, however, are somewhat larger in this case, because the measured amplitudes Δ_{per} (see Fig.2) are needed as input data. Note, that the seam Hamiltonian $\mathcal{H}_s$ is different from the one displayed in Eq.(2.2) for periodic boundary conditions.

Acknowledgments

Useful discussions with A. Ferrenberg and K.K. Mon are gratefully acknowledged. M. Krech gratefully acknowledges financial support through the Feodor Lynen program of the Alexander von Humboldt–Foundation. This research was supported in part by NSF grant # DMR–9405018.

References

[1] H.B.G. Casimir, Proc. K. Ned. Akad. Wet. **51**, 793 (1948); Physica **19**, 846 (1953).

[2] J. Schwinger, L.L. DeRaad, and K.A. Milton, Ann. Phys. (N.Y.) **115**, 1 (1978).

[3] I.E. Dzyaloshinskii, E.M. Lifshitz, and L.P. Pitaevskii, Adv. Phys. **10**, 165 (1961).

[4] R. Eisenschitz and F. London, Z. Phys **60**, 491 (1930).

[5] S. Dietrich, in *Phase Transitions and Critical Phenomena*, edited by C. Domb and J.L. Lebowitz (Academic, London, 1986), Vol.10, p.76.

[6] M. Krech and S. Dietrich, Phys. Rev. A **46**, 1886 and 1922 (1992).

[7] J.O. Indekeu, J. Chem. Soc., Faraday Trans. II **12**, 1835 (1986).

[8] M.P. Nightingale and J.O. Indekeu, Phys. Rev. Lett. **54**, 1824 (1985); **55**, 1700 (1985).

[9] R. Lipowsky and U. Seifert, Phys. Rev. B **31**, 4701 (1985); R. Lipowsky, Phys. Rev. Lett. **55**, 1699 (1985).

[10] J.L. Cardy, Nucl. Phys. **B275**, 200 (1985).

[11] J.O. Indekeu, M.P. Nightingale, and W.V. Wang, Phys. Rev. B **34**, 330 (1986).

[12] K.K. Mon, Phys. Rev. Lett. **54**, 2671 (1985); Phys. Rev. B **35**, 3560 (1987).

[13] V. Privman, P.C. Hohenberg, and A. Aharony, in *Phase Transitions and Critical Phenomena*, edited by C. Domb and J.L. Lebowitz (Academic, New York, 1991), Vol.14, p.1.

[14] A.M. Ferrenberg, D.P. Landau, and Y.J. Wong, Phys. Rev. Lett. **69**, 3382 (1992).

[15] A.M. Ferrenberg and R.H. Swendsen, Phys. Rev. Lett. **63**, 1195 (1989).

[16] A.E. Ferdinand and M.E. Fisher, Phys. Rev. **185**, 832 (1969).

[17] T.W. Burkhardt, private communication.

New Scheme of Hybrid Exponential Product Formulas with Applications to Quantum Monte-Carlo Simulations

M. Suzuki

Department of Physics, University of Tokyo, Bunkyo-ku, Tokyo 113, Japan

Abstract. The present paper reviews the hybrid scheme of exponential product formulas which includes some restricted number of commutators of the original operators. There are two types of hybrid schemes, namely, type A and type B.

1. Introduction

In the present paper, we propose a new scheme of exponential product formulas. The main difference from the ordinary decomposition is that not only the basic commutators but also some of their commutators are used. First we review briefly the standard product formulas of the form[1-5]

$$F_m(x) = e^{-t_1 A} e^{-t_2 B} e^{-t_3 A} e^{-t_4 B} \cdots e^{-t_M A}. \tag{1.1}$$

An exponential operator of the form $\exp[-x(A+B)]$ is approximated as

$$e^{-x(A+B)} = F_m(x) + O(x^{m+1}) \tag{1.2}$$

with $F_m(x)$ given by Eq.(1.1). It is well known[2,6] that some of the parameters $\{t_j\}$ in (1.1) are negative for $m \geq 3$ when $x > 0$. This causes the serious problem that $e^{-t_j A}$ can not be defined when $t_j < 0$ and when A is a positive unbounded operator. Thus, we try here to resolve this problem by adding some commutators of the operators A and B to the relevant product formulas. The above scheme using only A and B (not using commutators) will be called the standard product formula.

2. General Procedure for Constructing Hybrid Product Formulas

In the present section, we study a new scheme of product formulas including some commutators of A and B. Our motivation is to overcome the difficulty of appearance of negative parameters in (1.1) for $m \geq 3$, as already mentioned in Section 1. For this purpose, we extend here the basis space $(C_1, C_2, \cdots, C_r)$ in the expansion

$$e^{-x(A+B)} = e^{-C_1} e^{-C_2} \cdots e^{-C_r} + O(x^{m+1}), \tag{2.1}$$

to some restricted but slightly larger space of A, B and their commutators. If we extend it to the whole space, then such an expansion will be equivalent to the Zassenhaus formula

$$e^{-x(A+B)} = e^{-xA} e^{-xB} e^{-\frac{x^2}{2}[A,B]} e^{-\frac{x^3}{6}[A+2B,[A,B]]} \cdots. \tag{2.2}$$

Springer Proceedings in Physics, Vol. 80
Computer Simulation Studies in Condensed-Matter Physics VIII
Editors: D.P. Landau · K.K. Mon · H.-B. Schüttler

Now we rewrite the right-hand side of (2.1) in the form

$$e^{-C_1}e^{-C_2}\cdots e^{-C_r} = \exp\{-x(A+B) + x^2R_2(A,B) + x^3R_3(A,B) + \cdots\}. \tag{2.3}$$

Then, the required condition on $\{R_j(A,B)\}$ is given by

$$R_2(A,B) = 0,\ R_3(A,B) = 0, \cdots,\ R_m(A,B) = 0. \tag{2.4}$$

This determines the coefficients in (2.1).

For example, we may consider the following simple space of free Lie elements :

$$a)\ A, B \text{ and } [A,B] \text{ or } b)\ A, B \text{ and } [B,[A,B]]. \tag{2.5}$$

There may be another possibility of adding nonlinear terms of commutators such as A^2, B^2 and $[A,B]^2$ to the expansion space $(C_1, \cdots, C_r)$ in (2.1). However, we can prove the following statement[5].

Basic Theorem : *When an expenential operator* $e^{-x(A+B)}$ *is expressed in the form* (2.1), *each* C_j *is proportional to one of free Lie elements, namely* A, B *and commutators of* A *and* B. *Even if we add nonlinear terms, they do not effect the coefficients of free Lie elements in* (2.1).

This theorem is proved[5] using the Friedrichs theorem[7], as in proving the Baker-Hausdorff and Zassenhaus theorems[7].

Thus, we can confine our basis $(C_1, C_2, \cdots, C_r)$ into a linear space of free Lie elements, namely A, B and their commutators. As was generally shown by the present author[1-5], we do not need include any of commutators if the coefficients of A and B are allowed to take both positive or negative values.

The condition that all the coefficients of A and B in (2.1) should have the same sign requires including some commutators of the operators A and B. Then, our strategy of constructing a new scheme satisfying the above condition is the following. First, we consider a product of the form

$$G_N(x) = e^{-s_1A}e^{-s_2B}e^{-s_3A}e^{-s_4B}\cdots e^{-s_NA} \tag{2.6}$$

with $s_j > 0$ for all j. From our general theorem[1,2], the above product $G_N(x)$ can not be of order higher than the second and consequently we may write as

$$G_N(x) = e^{-x(A+B)+x^3R_3(A,B)+O(x^5)}, \tag{2.7}$$

when $s_{N-j+1} = s_j$ (symmetric). In general, the third-order correction $R_3(A,B)$ consists of the two commutators $[A,[A,B]]$ and $[B,[A,B]]$. We can choose the parameters $\{s_j\}$ in (2.6) so that one of the above two commutators may not appear in $R_3(A,B)$. For example, we consider here the situation in which only the commutator $[B,[A,B]]$ remains in $R_3(A,B)$. Then, we may put $G_N(x)$ as

$$G_N(x) = e^{-x(A+B)+2c_3x^3[B,[A,B]]+O(x^5)}. \tag{2.8}$$

Now, the hybrid fourth-order product formula is given by

$$S_4(x) = Q(x)G_N(x)Q(x)\ ;\ Q(x) = e^{-c_3x^3[B,[A,B]]}. \tag{2.9}$$

Alternatively we may start from

$$H_N(x) = S(p_1x)S(p_2x)\cdots S(p_Nx)\ ;\ \sum p_j = 1 \tag{2.10}$$

with $p_{N-j+1} = p_j > 0$ for all j, and with $S(x) = e^{-\frac{x}{2}A}e^{-xB}e^{-\frac{x}{2}A}$. The product $H_N(x)$ may be written in the form

$$H_N(x) = e^{-x(A+B)+x^3R_3'(A,B)+\cdots}. \tag{2.11}$$

The remaining procedure is the same as before. Thus, we find an alternative fourth-order scheme of the form

$$S_4'(x) = Q'(x)H_N(x)Q'(x) \;\; ; \;\; Q'(x) = e^{-\frac{x^3}{2}R_3'(A,B)}. \tag{2.12}$$

The correction term $R_3'(A,B)$ can be again chosen as $R_3'(A,B) = 2c_3'[B,[A,B]]$. Consequently we have

$$S_4'(x) = e^{-c_3'x^3[B,[A,B]]}S(p_1x)\cdots S(p_Nx)e^{-c_3'x^3[B,[A,B]]} \tag{2.13}$$

with $p_{N-j+1} = p_j > 0$ for all j.

Similarly, we consider a product of the form $J_N(x) = e^{-C_1}e^{-C_2}\ldots e^{-C_N}$, where C_j is proportional to one of A, B and $[B,[A,B]]$. Then, the coefficients of A, B and $[B,[A,B]]$ in $J_N(x)$ can be chosen to be all positive and yet to satisfy the requirement that $J_N(x)$ should be of the fourth order and symmetric. That is, we may write it as

$$J_N(x) = e^{-x(A+B)+x^5R_5(A,B)+O(x^7)}. \tag{2.14}$$

The correction $R_5(A,B)$ consists of the following six independent commutators[7] :

$$\begin{gathered}[[[[A,B],B],B],B], \;\; [[[[B,A],A],A],A], \;\; [[[[A,B],A],A],B], \\ [[[[A,B],A],B],B], \;\; [[[A,B],A],[A,B]], \;\; \text{and} \;\; [[[A,B],B],[A,B]]. \end{gathered} \tag{2.15}$$

The parameters $\{p_j\}$ in (2.10) can be chosen so that some desired number of the above six commutators should remain in $R_5(A,B)$. The above procedures can be repeated up to any higher order.

This is our general strategy for constructing the hybrid product scheme.

3. Some Examples of Hybrid Product Formulas

In the present section, we explain how to use the above strategy through the derivation of two examples of hybrid product formulas.

3.1 Hybrid scheme of type A

The first scheme is based only on A, B and $[A,B]$. It is easy[5] to confirm the formulas

$$e^{-\frac{x}{2}A}e^{-xB}e^{-\frac{x}{2}A} = e^{-x(A+B)+x^3[A+2B,C]+O(x^5)}, \tag{3.1}$$

$$e^{-x^2C}e^{-xB}e^{x^2C} = e^{-xB-x^3[B,C]+O(x^5)}, \tag{3.2}$$

and consequently

$$Q_2(x) \equiv e^{-\frac{x}{2}A}(e^{-x^2C}e^{-xB}e^{x^2C})e^{-\frac{x}{2}A} = e^{-x(A+B)+x^3[A+B,C]+O(x^5)}, \tag{3.3}$$

where $C = \frac{1}{24}[A,B]$. Therefore, we arrive at

$$Q_4(x,t) = e^{-tx^2C}Q_2(x)e^{tx^2C}$$
$$= \exp\{-x(A+B)+x^3(1-t)[A+B,C]+O(x^5)\}. \tag{3.4}$$

Therefore, $Q_4(x,1)$ yields a fourth-order hybrid product formula of type A, namely we obtain

$$Q_4(x) \equiv Q_4(x,1) = e^{-x^2C}e^{-\frac{x}{2}A}e^{-x^2C}e^{-xB}e^{x^2C}e^{-\frac{x}{2}A}e^{x^2C}. \tag{3.5}$$

This formula will be useful in studying a wavefunction in statistical mechanics.

3.2 Hybrid scheme of type B

There is another hybrid product formula of the fourth order[5]. It is based only on A, B and $[B,[A,B]]$, and is called "type B". It is easy[5] to derive the formula

$$S_{aba}(x) = S_a(px)S_b(qx)S_a(px)$$
$$= \exp\{-x(2p+q)(A+B)+x^3[2(p^3-q^3)A+(4p^3-q^3)B,C]+O(x^5)\}, \tag{3.6}$$

where $C=\frac{1}{24}[A,B]$, $S_a(x)=e^{-\frac{x}{2}A}e^{-xB}e^{-\frac{x}{2}A}$ and $S_b(x)=e^{-\frac{x}{2}B}e^{-xA}e^{-\frac{x}{2}A}$.
According to the general procedure in Section 2, we choose the parameters p and q so that the commutator $[A,C]=\frac{1}{24}[A,[A,B]]$ may vanish in (3.6). Putting $p=q=\frac{1}{3}$, we obtain

$$S_{aba}(x) = \exp\{-x(A+B)+\frac{x^3}{9}[B,C]+O(x^5)\}. \tag{3.7}$$

Then, a fourth-order hybrid product formula is given by[5]

$$S_4(x) = Q(x)S_{aba}(x)Q(x) = Q(x)S_a(\frac{x}{3})S_b(\frac{x}{3})S_a(\frac{x}{3})Q(x), \tag{3.8}$$

where

$$Q(x) = \exp(-\frac{1}{2}(\frac{x}{6})^3[B,[A,B]]). \tag{3.9}$$

Clearly, the product

$$S_4'(x) = S_a(\frac{x}{3})Q(x)S_b(\frac{x}{3})Q(x)S_a(\frac{x}{3}) \tag{3.10}$$

is also another fourth-order hybrid product formula[5].

These formulas are useful when we study electron systems described by the Hamiltonian $A=-\frac{1}{2}\Delta$ and $B=V(r)$, because we have $[B,[A,B]]=|\nabla V(r)|^2 \geq 0$ for the above operators A and B.

It should be remarked here that the above hybrid product formulas $S_4(x)$ and $S_4'(x)$ are contraction operators well defined for $x>0$, even when A and B are unbound positive hermitian operators.

4. Relation between the Hybrid Scheme and McLachlan's Method of Small Perturbation

The present hybrid scheme depends on how to choose the basis of its product expansion. A typical choice of the basis may be to make use of McLachlan's method of small perturbation[8]. Namely, we regard B as a small perturbation compared to A. Then, the commutator $[B,[A,B]]$ is of order higher than $[A,[A,B]]$ is. Consequently, we may write

$A+B$ as $A+\varepsilon B$ with a small parameter ε. As a simple example of $G_N(x)$ in (2.6), we consider

$$G(x,s) = e^{-sx\varepsilon B} e^{-\frac{x}{2}A} e^{-(1-2s)x\varepsilon B} e^{-\frac{x}{2}A} e^{-sx\varepsilon B}. \tag{4.1}$$

Following McLachlan[8], we have

$$\begin{aligned} G(x,s) &= \exp\{-x(A+\varepsilon B) + \frac{\varepsilon x^3}{24}(6s-1)[A,[B,A]] \\ &+ \frac{\varepsilon^2 x^3}{12}(6s^2-6s+1)[B,[A,B]] + O(x^5)\}. \end{aligned} \tag{4.2}$$

Then, we require[8] that the correction term of order ε in (4.2) should vanish. This yields the equation $6s-1=0$, namely $s=\frac{1}{6}$. Consequently we have

$$G(x,\frac{1}{6}) = \exp\{-x(A+\varepsilon B) + 2x^3(\frac{\varepsilon}{12})^2[B,[A,B]] + O(x^5)\}. \tag{4.3}$$

According to the present general procedure in Section 2, we find a fourth-order hybrid product formula of the form

$$Q_4'(x) = e^{-x^3 D} e^{-\frac{x}{6}B} e^{-\frac{x}{2}A} e^{-\frac{2}{3}xB} e^{-\frac{x}{2}A} e^{-\frac{x}{6}B} e^{-x^3 D}, \tag{4.4}$$

where we put $\varepsilon = 1$ and $D = \frac{1}{6}[B,C] = \frac{1}{144}[B,[A,B]]$.

By exchanging A and εB in (4.1), we may consider

$$G'(x,s) = e^{-sxA} e^{-\frac{x}{2}\varepsilon B} e^{-(1-2s)xA} e^{-\frac{x}{2}\varepsilon B} e^{-sxA}. \tag{4.5}$$

Then, we obtain the equation $6s^2-6s+1=0$ from the requirement that the correction term of order ε should vanish. Thus, we arrive at another fourth-order hybrid product formula

$$Q_4''(x) = e^{-x^3 D'} e^{-sxA} e^{-\frac{x}{2}B} e^{-(1-2s)xA} e^{-\frac{x}{2}B} e^{-sxA} e^{-x^3 D'} \tag{4.6}$$

where $D' = \frac{1}{48}(2-\sqrt{3})[B,[A,B]]$ and $s = \frac{3-\sqrt{3}}{6}$, i.e., $1-2s = \frac{\sqrt{3}}{3}$.

The above method can be extended to higher orders, as will be reported elsewhere.

5. Discussion

In the present paper, we have explained the general procedure for constructing hybrid product formulas and we have discussed the relation between the present hybrid product scheme and McLachlan's method of small perturbation. The present hybrid scheme will be very useful in quantum Monte Carlo simulations[9-18] of electron systems.

Acknowledgement

The author would like to thank H. Kobayashi for his useful comments.

References

1) M. Suzuki : Phys. Lett. **A146** (1990) 319 ; ibid **A165** (1992) 387.

2) M. Suzuki : J. Math. Phys. **32** (1991) 400.

3) M. Suzuki : Physica **A205** (1994) 65, and references cited therein.

4) M. Suzuki : Commun. Math. Phys. **163** (1994) 491.

5) M. Suzuki : Phys. Lett. A (1995) (in press).

6) Q. Sheng : IMA J. Numer. Anal. **9** (1989) 199.

7) W. Magnus, A. Karrass and D. Solitar : *Combinatorial Group Theory* (Dover, 1976).

8) R.I. McLachlan : Composition methods in the presence of small parameters (preprint).

9) M. Suzuki : Prog. Theor. Phys. **56** (1976) 1454.

10) M. Suzuki, S. Miyashita and A. Kuroda : Prog. Thoer. Phys. **58** (1977) 1377.

11) M.H. Kalos : ed., *Monte Carlo Methods in Quantum Problems* (Reidel, Boston, 1976).

12) H. De Raedt and A. Lagendijk : Phys. Rep. **127** (1985) 233.

13) M. Suzuki : J. Math. Phys. **26** (1985) 601.

14) M. Suzuki : J. Stat. Phys. **43** (1986) 883.

15) M. Suzuki : ed., *Quantum Monte Carlo Methods,* Solid State Sciences, vol.74 (Springer, Berlin, 1986).

16) M. Suzuki : J. Stat. Phys. **49** (1987) 977.

17) M. Suzuki : Physica A **194** (1993) 432.

18) M. Suzuki : ed., *Quantum Monte Carlo Methods in Condensed Matter Physics* (World Scientific, Singapore, 1993), and references cited therein.

Spin Dynamics Calculations in the Two-Dimensional Classical XY-Model

H.G. Evertz[1] *and D.P. Landau*[2]

Center for Simulational Physics, University of Georgia,
Athens, GA 30602, USA
[1]evertz@csp2.csp.uga.edu
[2]dlandau@uga.cc.uga.edu

Abstract. We report preliminary results from the first large scale numerical study of critical spin dynamics in the two dimensional classical XY model. We integrate the hamiltonian equations of motion, starting from a set of configurations generated by Monte Carlo. In addition to the expected spin wave peak in the Kosterlitz-Thouless phase, we find a strong central peak, and unexpected structure below the spin wave peak. The dynamic critical exponent is measured to be $z = 1.00(4)$ in the KT-phase.

1 Introduction

We have investigated the dynamic critical behavior of the two-dimensional classical anisotropic Heisenberg model (XY-model)

$$\mathcal{H} = -J \sum_{nn} \left(\mathbf{S}_i^x \mathbf{S}_j^x + \mathbf{S}_i^y \mathbf{S}_j^y + \lambda \mathbf{S}_i^z \mathbf{S}_j^z \right) \quad , \lambda \equiv 0 \,. \tag{1}$$

The statics of this model are similar to those of the "plane rotator model", i.e. the model with *two*-component spins: at all temperatures $T \leq T_{KT}$ the model is in a Kosterlitz-Thouless phase [1], where it is dominated by vortex-pairs, does not have long range order, and is *critical*, with static correlations decaying like $\langle S(0)S(r) \rangle \sim r^{-\eta}$. The static critical exponent η is 1/4 at T_{KT}. Above T_{KT}, vortices unbind and the correlations decay exponentially.

The *dynamics*, however, are very different: The XY-model possesses equations of motion

$$\frac{d}{dt}\mathbf{S}_i = \mathbf{S}_i \times \left[-J \sum_{nn} \left(\mathbf{S}_j^x \hat{\mathbf{e}}_x + \mathbf{S}_j^y \hat{\mathbf{e}}_y \right) \right] , \tag{2}$$

where $\hat{\mathbf{e}}_x$ and $\hat{\mathbf{e}}_y$ are unit vectors in the x- and y-directions respectively. Eq. (2) is a set of coupled equations and can be integrated numerically. The plane rotator model, on the other hand, does not possess hamiltonian equations of motion; there is only relaxational dynamics.

To obtain the dynamic critical behavior for each temperature, we generated a set of equilibrium configurations by Monte Carlo simulations and integrated the hamiltonian equations of motion for each configuration up to

Computer Simulation Studies in Condensed-Matter Physics VIII
Editors: D.P. Landau · K.K. Mon · H.-B. Schüttler

a maximum time t_{max}. For each time evolution, we measured the time-displaced, space-displaced spin-spin correlation function. The time- and space-fourier transform of this function is the dynamic structure factor (neutron scattering function) $S(q,\omega)$.

In the present paper we emphasize several technical aspects of our study, as well as an unexpected structure observed in $S(q,\omega)$. We will present a detailed physical discussion of $S(q,\omega)$ elsewhere [2]. An earlier exploratory study [3] indicated a rich structure in the neutron scattering function which was not adequately described by theory.

2 Simulations and Time Integration

Most of our simulations were performed below T_{KT}, where the system is critical. The spatial correlation length there is only limited by the system size. Standard Monte Carlo procedures will therefore suffer from severe Critical Slowing Down. In order to reduce autocorrelations we employed a hybrid method in which each update consisted of 2 fully vectorized checkerboard Metropolis updates, 8 Overrelaxation updates [4], and one Single-Cluster update [5]. In the overrelaxation algorithm [4] each spin is reflected with respect to a plane, in such a way that its contribution to the total energy remains constant. The spin z-component is not changed during this update. The algorithm is vectorized in checkerboard fashion. The cluster algorithm is also restricted to changes in the xy-plane. A single cluster is constructed, in complete analogy with Wolff's cluster algorithm [5] for the plane rotator model. Since both overrelaxation and Cluster updates leave the z-components of spins unchanged, Metropolis updates are necessary to ensure ergodicity. Inclusion of the cluster updates reduced autocorrelations drastically, e.g. at $T = 0.6$, $L = 128$, from about 300 hybrid sweeps when omitting the cluster updates down to about 3 hybrid sweeps.

We generated between 500 and 1200 independent spin-configurations for each combination of temperatures $T = 0.4$, 0.6, $0.725 \approx T_{KT}$, $T = 0.8$ and lattice sizes from 16^2 to 192^2, with about 200 hybrid sweeps between configurations. We found this many configurations to be necessary in order to sufficiently reduce thermal fluctuations in the resulting neutron scattering function. The error bars in our figures represent the statistical errors for averages over the equilibrium configurations.

Starting with each equilibrium configuration, the time dependence of the spins was determined from the coupled set of equations of motion, eq. (2), which was integrated numerically using a vectorized fourth order predictor-corrector method [6], with a time step size of $\delta t = 0.01J^{-1}$. This method has a very small systematical error, of order $(\delta t)^5$. The maximum integration time was $t_{max} = 400J^{-1}$. (A few runs were also performed for lattice size 256^2 with $t_{max} = 800J^{-1}$ and produced the same physical results.) We chose a very large t_{max} in order to sufficiently reduce cutoff effects in $S(q,\omega)$, as discussed below. This value of t_{max} is much larger than any employed in previous studies, and the time integration could potentially become unstable. We checked stability by verifying that $S(q,\omega)$ remained unchanged when we introduced an *additional* time integration of length $200J^{-1}$ before starting to measure space-time correlation functions. In addition, we found that the total energy changed by a relative factor of less than 3×10^{-6} between $t=0$ and t_{max}.

3 Extracting $S(q,\omega)$

In order to determine $S(q,\omega)$, we measured the space displaced, time displaced spin-spin correlation function $C^{kk}(\mathbf{r}-\mathbf{r}',t-t') = \langle S^k_{\mathbf{r}}(t)S^k_{\mathbf{r}'}(t')\rangle$ for each time evolution, with $t-t' \le t_{cut} \equiv 0.9\,t_{max}$, and averaged results over the different time evolutions, and over $t\in\{0,t_{max}/10\}$. Fourier transforms in space and time then gave $S^{kk}(\vec{q},\omega)$. Spatial xy symmetry allows averaging over S^{xx} and S^{yy}. We saved a large amount of storage space by fixing the direction of $\vec{q}$ to $(q,0)$ or $(0,q)$, and averaging over the results. Since $S(q,\omega)$ is a convolution, it can be computed efficiently by use of Fast Fourier Transforms, which saved a large amount of CPU time.

Note that a temporal fourier transform with a finite time cutoff t_{cut} produces *oscillations* in $S(q,\omega)$, with period $2\pi/t_{cut}$. These oscillations hinder analysis of $S(q,\omega)$, for example at the very narrow spin wave peak. To reduce the impact of the cutoff, one can *smoothen* $S(q,\omega)$ by multiplying the correlation function with a damping factor, e.g. $\exp(-\frac{1}{2}(t\,\delta\omega)^2)$. Smoothening widens all features in $S(q,\omega)$, similar to the effect of finite ω-resolution in an

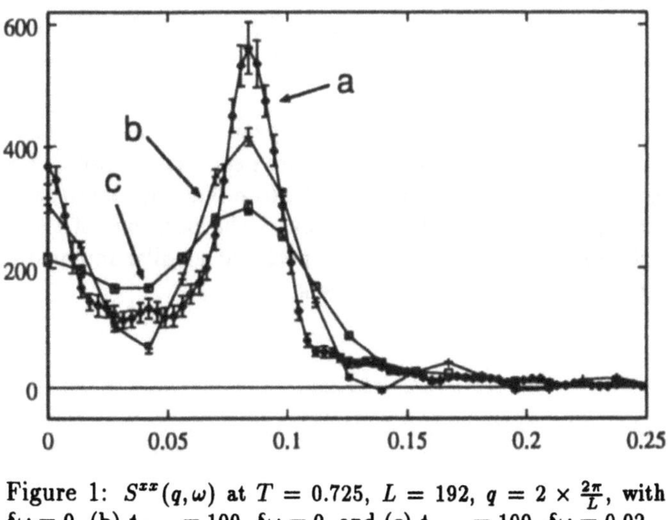

Figure 1: $S^{xx}(q,\omega)$ at $T=0.725$, $L=192$, $q=2\times\frac{2\pi}{L}$, with (a) $t_{max}=400$, $\delta\omega=0$, (b) $t_{max}=100$, $\delta\omega=0$, and (c) $t_{max}=100$, $\delta\omega=0.02$.

experiment. It also complicates analysis of the data by Dynamic Finite Size Scaling [6].

Fig. 1 shows an example of $S(q,\omega)$ with different values of t_{max}. For small $t_{max} = 100J^{-1}$, oscillations are strong, and the spin wave peak is too broad. A damping factor of $\delta\omega = 0.02$ smoothens the oscillations, but it also further broadens the peak. Both oscillations and broadening are overcome by integrating up to $t_{max} = 400J^{-1}$. For critical dynamics, larger lattice sizes necessitate larger integration times. We found dynamic finite size scaling to be a very sensitive tool to determine which values of t_{max} are sufficient [2].

4 Results

We summarize briefly the main conclusions from a first analysis of our data. Detailed results will be presented elsewhere [2].

The two main physical features in $S(q,\omega)$ are the spin wave peak and an unexpectedly large central peak around $\omega = 0$. The spin wave peak is present in S^{xx} at all temperatures $T \leq T_{KT}$; above T_{KT} there is only a large central peak. As T increases, the spin wave peak in S^{xx} becomes wider and moves to smaller ω; with increasing q it becomes wider and weaker. Its shape is not well described by existing theoretical calculations [7]. There is a sizeable central peak in S^{xx} even at $T \leq T_{KT}$, which was not predicted. It becomes wider and stronger with increasing T. The out-of-plane correlations S^{zz} also show a weak and rather sharp spin wave peak at all temperatures. It is wider than the predicted delta function [7] would produce with our finite integration time. The dispersion relation (position of the spin wave peak versus momentum) is linear at small q for $T \leq T_{KT}$, and nonlinear above T_{KT}, as expected.

By dynamic finite size scaling [2] we determined the dynamic critical exponent z. Both a characteristic frequency ω_m and $S(q,\omega)$ itself scale very well, and we obtained $z = 1.00(4)$ for all $T \leq T_{KT}$, in agreement with theoretical predictions [7].

In addition to the spin wave peak and central peak, there is additional unexpected structure in S^{xx}, namely several smaller peaks at multiples of a basic frequency ω_b. An example is shown in fig. 2. Note that the logarithmic scale strongly overemphasizes the additional peaks. The positions of these peaks coincide with the positions of spin-wave peaks at smaller momenta. Analysis of this intriguing structure is in progress.

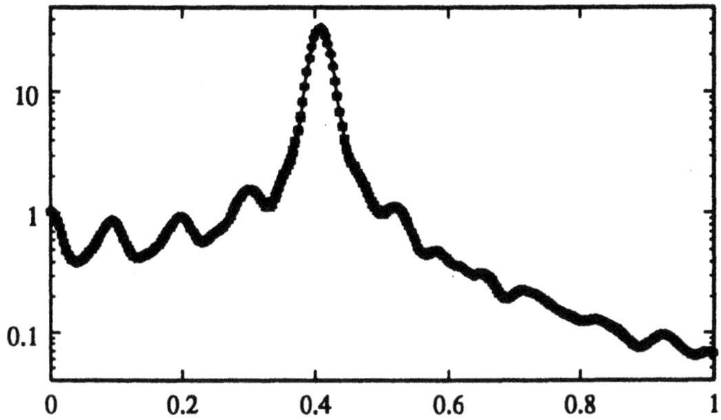

Figure 2: $S^{xx}(q,\omega)$ at $T = 0.600$, $L = 192$, $q = 8 \times \frac{2\pi}{L}$.

This research was supported by NSF Grant # DMR-9405018. Calculations were done on the Cray C90 at the Pittsburgh Supercomputing Center.

References

[1] J. M. Kosterlitz and D. J. Thouless, J. Phys. C **6**, 1181 (1973).

[2] H. G. Evertz and D. P. Landau, in preparation.

[3] D. P. Landau and R. W. Gerling, J. Magn. Magn. Mat. **104-107**, 843 (1992).

[4] F. R. Brown and T. J. Woch, Phys. Rev. Lett. **58**, 2394 (1987); M. Creutz, Phys. Rev. D **36**, 515 (1987).

[5] U. Wolff, Phys. Rev. Lett. **62**, 361 (1989), Nucl. Phys. B **322**, 759 (1989), and Phys. Lett. **228B**, 379 (1989).

[6] K. Chen and D. P. Landau, Phys. Rev. B **49**, 3266 (1994).

[7] J. Villain, J. Phys. (Paris) **35**, 27 (1974); D. R. Nelson and D. S. Fisher, Phys. Rev. B **16**, 4945 (1977); S. L. Menezes, A. S. T. Pires, and M. E. Gouvêa, Phys. Rev. B **47**, 12280 (1993), Phys. Rev. B **45**, 10454 (1992); A. R. Pereira, A. S. T. Pires, M. E. Gouvêa, and B. V. Costa, Z. Phys. B. **89**, 109 (1992).

Index of Contributors

Springer Proceedings in Physics

Managing Editor: H. K. V. Lotsch

1 *Fluctuations and Sensitivity in Nonequilibrium Systems*
Editors: W. Horsthemke and D. K. Kondepudi
2 *EXAFS and Near Edge Structure III*
Editors: K. O. Hodgson, B. Hedman, and J. E. Penner-Hahn
3 *Nonlinear Phenomena in Physics*
Editor: F. Claro
4 *Time-Resolved Vibrational Spectroscopy*
Editors: A. Laubereau and M. Stockburger
5 *Physics of Finely Divided Matter*
Editors: N. Boccara and M. Daoud
6 *Aerogels* Editor: J. Fricke
7 *Nonlinear Optics: Materials and Devices*
Editors: C. Flytzanis and J. L. Oudar
8 *Optical Bistability III*
Editors: H. M. Gibbs, P. Mandel, N. Peyghambarian, and S. D. Smith
9 *Ion Formation from Organic Solids (IFOS III)*
Editor: A. Benninghoven
10 *Atomic Transport and Defects in Metals by Neutron Scattering*
Editors: C. Janot, W. Petry, D. Richter, and T. Springer
11 *Biophysical Effects of Steady Magnetic Fields*
Editors: G. Maret, I. Kiepenheuer, and N. Boccara
12 *Quantum Optics IV*
Editors: J. D. Harvey and D. F. Walls
13 *The Physics and Fabrication of Microstructures and Microdevices*
Editors: M. J. Kelly and C. Weisbuch
14 *Magnetic Properties of Low-Dimensional Systems*
Editors: L. M. Falicov and J. L. Morán-López
15 *Gas Flow and Chemical Lasers*
Editor: S. Rosenwaks
16 *Photons and Continuum States of Atoms and Molecules*
Editors: N. K. Rahman, C. Guidotti, and M. Allegrini
17 *Quantum Aspects of Molecular Motions in Solids*
Editors: A. Heidemann, A. Magerl, M. Prager, D. Richter, and T. Springer
18 *Electro-optic and Photorefractive Materials*
Editor: P. Günter
19 *Lasers and Synergetics*
Editors: R. Graham and A. Wunderlin
20 *Primary Processes in Photobiology*
Editor: T. Kobayashi
21 *Physics of Amphiphilic Layers*
Editors: J. Meunier, D. Langevin, and N. Boccara
22 *Semiconductor Interfaces: Formation and Properties*
Editors: G. Le Lay, J. Derrien, and N. Boccara
23 *Magnetic Excitations and Fluctuations II*
Editors: U. Balucani, S. W. Lovesey, M. G. Rasetti, and V. Tognetti
24 *Recent Topics in Theoretical Physics*
Editor: H. Takayama
25 *Excitons in Confined Systems*
Editors: R. Del Sole, A. D'Andrea, and A. Lapiccirella
26 *The Elementary Structure of Matter*
Editors: J.-M. Richard, E. Aslanides, and N. Boccara
27 *Competing Interactions and Microstructures: Statics and Dynamics*
Editors: R. LeSar, A. Bishop, and R. Heffner
28 *Anderson Localization*
Editors: T. Ando and H. Fukuyama
29 *Polymer Motion in Dense Systems*
Editors: D. Richter and T. Springer
30 *Short-Wavelength Lasers and Their Applications*
Editor: C. Yamanaka
31 *Quantum String Theory*
Editors: N. Kawamoto and T. Kugo
32 *Universalities in Condensed-Matter*
Editors: R. Jullien, L. Peliti, R. Rammal, and N. Boccara
33 *Computer Simulation Studies in Condensed-Matter Physics: Recent Developments*
Editors: D. P. Landau, K. K. Mon, and H.-B. Schüttler
34 *Amorphous and Crystalline Silicon Carbide and Related Materials*
Editors: G. L. Harris and C. Y.-W. Yang
35 *Polycrystalline Semiconductors: Grain Boundaries and Interfaces*
Editors: H. J. Möller, H. P. Strunk, and J. H. Werner
36 *Nonlinear Optics of Organics and Semiconductors*
Editor: T. Kobayashi
37 *Dynamics of Disordered Materials*
Editors: D. Richter, A. J. Dianoux, W. Petry, and J. Teixeira
38 *Electroluminescence*
Editors: S. Shionoya and H. Kobayashi
39 *Disorder and Nonlinearity*
Editors: A. R. Bishop, D. K. Campbell, and S. Pnevmatikos
40 *Static and Dynamic Properties of Liquids*
Editors: M. Davidović and A. K. Soper
41 *Quantum Optics V*
Editors: J. D. Harvey and D. F. Walls
42 *Molecular Basis of Polymer Networks*
Editors: A. Baumgärtner and C. E. Picot
43 *Amorphous and Crystalline Silicon Carbide II: Recent Developments*
Editors: M. M. Rahman, C. Y.-W. Yang, and G. L. Harris
44 *Optical Fiber Sensors*
Editors: H. J. Arditty, J. P. Dakin, and R. Th. Kersten
45 *Computer Simulation Studies in Condensed-Matter Physics II: New Directions*
Editors: D. P. Landau, K. K. Mon, and H.-B. Schüttler

MIX
Papier aus verantwortungsvollen Quellen
Paper from responsible sources
FSC® C105338

If you have any concerns about our products,
you can contact us on
ProductSafety@springernature.com

In case Publisher is established outside the EU,
the EU authorized representative is:
Springer Nature Customer Service Center GmbH
Europaplatz 3, 69115 Heidelberg, Germany

Printed by Libri Plureos GmbH
in Hamburg, Germany